VOLUME THREE HUNDRED AND SEVENTY ONE

International Review of CELL AND MOLECULAR BIOLOGY

Cellular Immunity in the Peritoneum

INTERNATIONAL REVIEW OF CELL AND MOLECULAR BIOLOGY

VOLUME THREE HUNDRED AND SEVENTY ONE

INTERNATIONAL REVIEW OF CELL AND MOLECULAR BIOLOGY

Cellular Immunity in the Peritoneum

Edited by

FERNANDO ARANDA
Cima Universidad de Navarra, Pamplona, Spain

PEDRO BERRAONDO
Cima Universidad de Navarra, Pamplona, Spain

LORENZO GALLUZZI
Weill Cornell Medical College,
New York, NY, United States

ELSEVIER

ACADEMIC PRESS
An imprint of Elsevier

Academic Press is an imprint of Elsevier
50 Hampshire Street, 5th Floor, Cambridge, MA 02139, United States
525 B Street, Suite 1650, San Diego, CA 92101, United States
The Boulevard, Langford Lane, Kidlington, Oxford OX5 1GB, United Kingdom
125 London Wall, London, EC2Y 5AS, United Kingdom

First edition 2022

ISBN: 978-0-323-99400-2
ISSN: 1937-6448

For information on all Academic Press publications
visit our website at https://www.elsevier.com/books-and-journals

Publisher: Zoe Kruze
Acquisitions Editor: Leticia M. Lima
Developmental Editor: Jhon Michael Peñano
Production Project Manager: Vijayaraj Purushothaman
Cover Designer: Vicky Pearson Esser

Typeset by STRAIVE, India

Contents

Contributors

Fernando Aranda
Program of Immunology and Immunotherapy, Cima Universidad de Navarra; Navarra Institute for Health Research (IdiSNA), Pamplona, Spain

Leire Arrizabalaga
Program of Immunology and Immunotherapy, Cima Universidad de Navarra; Navarra Institute for Health Research (IdiSNA), Pamplona, Spain

Ángela Bella
Program of Immunology and Immunotherapy, Cima Universidad de Navarra; Navarra Institute for Health Research (IdiSNA), Pamplona, Spain

Pedro Berraondo
Program of Immunology and Immunotherapy, Cima Universidad de Navarra; Navarra Institute for Health Research (IdiSNA), Pamplona; Centro de Investigación Biomédica en Red de Cáncer (CIBERONC), Madrid, Spain

Laura Carrillo-Serradell
Institut d'Investigacions Biomèdiques August Pi i Sunyer (IDIBAPS), Barcelona, Spain

Sergi Casadó-Llombart
Institut d'Investigacions Biomèdiques August Pi i Sunyer (IDIBAPS), Barcelona, Spain

Cristina Català
Institut d'Investigacions Biomèdiques August Pi i Sunyer (IDIBAPS), Barcelona, Spain

Wim Ceelen
Experimental Surgery Lab, Department of Human Structure and Repair, Faculty of Medicine and Health Sciences, Ghent University; Cancer Research Institute Ghent (CRIG), Ghent, Belgium

Sarah Cosyns
Experimental Surgery Lab, Department of Human Structure and Repair, Faculty of Medicine and Health Sciences, Ghent University; Cancer Research Institute Ghent (CRIG), Ghent, Belgium

Jesse Demuytere
Experimental Surgery Lab, Department of Human Structure and Repair, Faculty of Medicine and Health Sciences, Ghent University; Cancer Research Institute Ghent (CRIG), Ghent, Belgium

Claudia Augusta Di Trani
Program of Immunology and Immunotherapy, Cima Universidad de Navarra; Navarra Institute for Health Research (IdiSNA), Pamplona, Spain

Sam Ernst
Experimental Surgery Lab, Department of Human Structure and Repair; Laboratory of Experimental Cancer Research, Department of Human Structure and Repair, Faculty of

Medicine and Health Sciences, Ghent University; Cancer Research Institute Ghent (CRIG), Ghent, Belgium

Myriam Fernández-Sendin
Program of Immunology and Immunotherapy, Cima Universidad de Navarra; Navarra Institute for Health Research (IdiSNA), Pamplona, Spain

Caterina Fiorentini
Department of Prevention and Sport Medicine, University Hospital Klinikum rechts der Isar, Technical University of Munich, Munich, Germany

Giammaria Fiorentini
Department of Onco-Hematology, Azienda Ospedaliera "Ospedali Riuniti Marche Nord", Pesaro, Italy

Stefano Guadagni
Department of Applied Clinical Sciences and Biotechnology, University of L'Aquila, L'Aquila, Italy

Mirja Hommel
Centro de Investigación Médica Aplicada (CIMA), Universidad de Navarra; IdiSNA, Instituto de Investigación Sanitaria de Navarra, Pamplona, Spain

Marcos Isamat
Sepsia Therapeutics S.L. 08908 L'Hospitalet de Llobregat, Barcelona, Spain

Alejandra Leyton-Pereira
Institut d'Investigacions Biomèdiques August Pi i Sunyer (IDIBAPS), Barcelona, Spain

Diana Llopiz
Centro de Investigación Médica Aplicada (CIMA), Universidad de Navarra; IdiSNA, Instituto de Investigación Sanitaria de Navarra; CIBEREHD, Pamplona, Spain

Francisco Lozano
Institut d'Investigacions Biomèdiques August Pi i Sunyer (IDIBAPS); Servei d'Immunologia, Centre de Diagnòstic Biomèdic (CDB), Hospital Clínic de Barcelona; Departament de Biomedicina, Facultat de Medicina, Universitat de Barcelona, Barcelona, Spain

Ignacio Melero
Program of Immunology and Immunotherapy; Translational Oncology Group, Program in Solid Tumors, Cima Universidad de Navarra; Navarra Institute for Health Research (IdiSNA); Department of Immunology and Immunotherapy, Clinica Universidad de Navarra, Pamplona; Centro de Investigación Biomédica en Red de Cáncer (CIBERONC), Madrid, Spain

Flor Navarro
Centro de Investigación Médica Aplicada (CIMA), Universidad de Navarra; IdiSNA, Instituto de Investigación Sanitaria de Navarra, Pamplona, Spain

Sanjay Rathod
Department of Immunology, University of Pittsburgh, Pittsburgh, PA, United States

David Repáraz
Centro de Investigación Médica Aplicada (CIMA), Universidad de Navarra; IdiSNA, Instituto de Investigación Sanitaria de Navarra; CIBEREHD, Pamplona, Spain

Joan Salvador Russo-Cabrera
Program of Immunology and Immunotherapy, Cima Universidad de Navarra; Navarra Institute for Health Research (IdiSNA), Pamplona, Spain

Donatella Sarti
Oncology Department, S. Maria Della Misericordia Hospital, ASUR1, Urbino, Italy

Alvaro Teijeira
Program of Immunology and Immunotherapy, Cima Universidad de Navarra, Pamplona; Centro de Investigación Biomédica en Red de Cáncer (CIBERONC), Madrid, Spain

Judith van Ovost
Experimental Surgery Lab, Department of Human Structure and Repair, Faculty of Medicine and Health Sciences, Ghent University; Cancer Research Institute Ghent (CRIG), Ghent, Belgium

María Velasco-de Andrés
Institut d'Investigacions Biomèdiques August Pi i Sunyer (IDIBAPS), Barcelona, Spain

CHAPTER ONE

The role of dendritic cells in the immune niche of the peritoneum

David Repáraz[a,b,c,*], Mirja Hommel[a,b], Flor Navarro[a,b], and Diana Llopiz[a,b,c,*]

[a]Centro de Investigación Médica Aplicada (CIMA), Universidad de Navarra, Pamplona, Spain
[b]IdiSNA, Instituto de Investigación Sanitaria de Navarra, Pamplona, Spain
[c]CIBEREHD, Pamplona, Spain
[*]Corresponding authors: e-mail address: dreparaz@alumni.unav.es; diallo@unav.es

Contents

Abstract

Dendritic cells (DCs) are professional antigen presenting cells that play an important role in the induction of T cell responses. Different subsets (cDC1s, cDC2s, pDCs, and moDCs) were described based on the expression of different surface markers and functions. In the context of peritoneum, DCs are also a key population cell orchestrating immune responses against pathogens, malignant cells and tissue-damage. Furthermore, they play an important role in the promotion of an anti-inflammatory microenvironment, which is necessary to maintain tolerance and adipocyte homeostasis. The aim of this review is to summarize the current knowledge of the functional and phenotypic features of peritoneal DCs and shed some light on the importance of these cells within this unique cavity and its associated components: the omentum, the mesentery and gut-associated lymphoid tissue (GALT).

Abbreviations

APC antigen presenting cells
cDC conventional Dendritic Cell
DC dendritic cell
GALT gut associated lymphoid tissue

International Review of Cell and Molecular Biology, Volume 371
ISSN 1937-6448
https://doi.org/10.1016/bs.ircmb.2022.04.012

LPM large peritoneal macrophages
MLN mesenteric lymph nodes
moDC monocyte derived Dendritic Cell
MS milky spot
pDC plasmacytoid Dendritic Cell
SPM small peritoneal macrophages

1. Functions and general features of dendritic cells

Discovered by Steinman and Cohn in 1973 (Steinman and Cohn, 1973) dendritic cells (DCs) are the most efficient antigen presenting cells (APC). They play an important role in connecting innate and adaptive responses and orchestrating and modulating the action of both (Worbs et al., 2017), which leads to the activation of the immune response or to the induction of immune tolerance, depending on the circumstances and microenvironment (Banchereau et al., 2000).

DCs derive from hematopoietic bone marrow progenitors called common dendritic cell progenitors (Gardner et al., 2020), which migrate to peripheral tissues where they stay as immature DCs (imDCs) (Gardner et al., 2020; Liu et al., 2021a). imDCs are characterized by high phagocytic and processing activities (Banchereau et al., 2000; Gardner et al., 2020), enabling them to efficiently capture and process antigens. When this happens in the presence of pathogen- associated molecular patterns (PAMPs) and damage-associated molecular patterns (DAMPs) (infection and damage, respectively), their properties and surface markers change. They upregulate CCR7 and migrate to secondary lymphoid organs, which are enriched in the cognate ligands of CCR7, chemokines CCL19 and CCL21 (Liu et al., 2021a). Here they present antigens to naïve lymphocytes and activate them using the following three signals: (1) antigen-specific T-cell receptor (TCR) signaling induced by the interaction between the TCR and its specific MHC- antigen complex, (2) the interaction between the costimulatory molecules expressed by the DCs (CD80 and CD86, both upregulated after maturation) and CD28 on lymphocytes, and finally (3) cytokine-induced signaling, which further polarizes T cells toward a specific phenotype (Kambayashi and Laufer, 2014; Reis and Sousa, 2006). All three signals are required for efficient T cell activation and the lack of any of them could result in differentiation of T cells into alternative phenotypes. For instance, an insufficient signal 2 can result in T cell anergy and/or the induction of tolerance by the generation of T regulatory cells (Treg) (Fucikova et al., 2019; Hasegawa and Matsumoto, 2018).

2. The different subsets of dendritic cells

DCs are a heterogeneous cell population with specialized functions, which allows them to modulate and fine-tune the immune response to be more efficient and precise (Wculek et al., 2020). They can be divided in classical or conventional DCs (cDC), plasmacytoid DC (pDC), and monocyte- derived DCs (moDC). These DC subsets display distinct molecular patterns on their surface, depending not only on the subset they belong to, but also on their origin (see Table 1), and the tissue where they are located (Sohn et al., 2021). Additionally, there is a fourth DC subset known as Langerhans cells, which is present in the basal epidermis and stratified epithelia and has the ability to self-renew (Collin and Bigley, 2018; Liu et al., 2021b).

cDC derive from bone marrow precursors and they are usually identified as MHC-II- and CD11c-positive cells. They leave the bone marrow as pre-cDCs ($CD38^+$ $CD45RA^+$) and infiltrate peripheral tissues, where they become cDCs (Collin and Bigley, 2018; Guilliams et al., 2014). Morphologically, they have a stellate shape with extensive membrane protrusions, also known as dendrites (Wculek et al., 2020). These cDCs can be further divided into two subsets: cDC1 ($CD24^+$ and $XCR1^+$ in mice and $CD141^+$ in humans) and cDC2 ($CD172\alpha^+$ and $CD1c^+$ both in mice and humans). In the presence of type-I interferon (IFN), cDC1 perform very efficient cross-presentation and tend to induce Th1 responses, thereby promoting a robust proliferation and activation of $CD8^+$ T cells (Gardner et al., 2020; Murphy and Murphy, 2021). Thus, they play an important role in responses against intracellular pathogens and cancer. They can produce high quantities of IL-12 and type-I and type-III IFN and, in a steady state, they also orchestrate self-tolerance (Wculek et al., 2020). Depending on their localization, cDC1s display different surface markers, such as CD8α, which is expressed on lymphoid residents, or CD103, which is a marker on migratory cDC1s (Gardner et al., 2020; Sun et al., 2020; Wculek et al., 2020). On the other hand, cDC2s conversely polarize CD4 cells to Th2 and Th17 phenotypes (Gardner et al., 2020), generating responses against extracellular pathogens (Murphy and Murphy, 2021; Wculek et al., 2020). They also play an important role in the maintenance of the Treg population

Table 1 Overview of the different molecular patterns expressed by human and mouse dendritic cell (DC) subsets.

DCs		Surface markers: Human (shared)	Human	Mice (shared)	Mice	TF
cDC	cDC1	**HLA-DR** **CD11c** CD33 CD13	**CD141** (BDCA-3) **CADM1** (NECL2) **CD369** (CLEC9A) XCR1 CD226 (DNAM-1) CD26	**MHC-II** **CD11c**	**XCR1** **CD24** CD369 (CLEC9A) CADM1 BTLA	**IRF8 hi** **Batf3** **ID2** GATA2 NOTCH
	cDC2		**CD1c** (BDCA-1) **CD172a** (Sirpα) **CD301** (CLEC10A) CD11c CD11b CD367 (CLEC4A) CD371 (CLEC12A)		**CD1c** ***CD172α*** (SIRPα) CD11b F4/80	**IRF4** **ZEB2** ID2 NOTCH KLF4
pDC		**CD123** (IL3R) **CD303** (BDCA-2, CLEC4C) **CD304** (BDCA-4) CD45RA FCER1 CD272 (BTLA) CD85k (ILT3) CD85g (ILT7) CD358 (DR6/TNFRSF21)		**CD45R** (B220) CD11c Ly6C CD317 (PDCA1)		**E2–2** (TCF4) int **IRF8** int **IRF4** IKZF1 SPIB ZEB2 GATA2
moDC		**CD14** **CD64** (FcγRI) CD1c FCER1		**Ly6C** **CD64** (FcγRI) FCER1		IRF4 BLIMP1 AHR ZBTB46
LC		**CD207** (Langerin) **CD1a** *E*-CADHERIN EpCAM (TROP1) TROP2		**CD207** (Langerin) **CD24** F4/80 CD11c MHC-II		ID2 RUNX3

Most commonly used markers appear in bold letters (Chen et al., 2021; Gardner et al., 2020; Guilliams et al., 2014; Guilliams et al., 2016; Macri et al., 2018; Villar and Segura, 2020; Wculek et al., 2020). cDCs, conventional DCs; pDCs, plasmacytoid DCs; moDCs, monocytes DCs; LC, Langerhans Cells; and TF, transcription factor.

(Sun et al., 2020) and in the induction of CD8 responses by CD4 cell DC licensing via interactions between CD40L (CD154) on the T cell and CD40 on the DC (Smith et al., 2004; Thaiss et al., 2011). Subsequently DCs upregulate MHC class I and CD80/86, thus favoring CD8 T cell activation (Thaiss et al., 2011). In contrast to cDC1s, cDC2s express CD4 and CD11b in lymphoid tissues, and CD11b alone or in combination with CD103 when found in peripheral tissues (Sun et al., 2020).

The subset of pDCs arises from both lymphoid and common DC progenitors (Mccully and Madrenas, 2006) and expresses CD45R (B220) in mice and CD123 in humans, among other molecules. These DCs play important roles in the clearance of viral agents due to their high production of inflammatory cytokines, such as type-I IFN, IL-6, and TNF-α (Collin and Bigley, 2018; Gardner et al., 2020; Macri et al., 2018). In addition, they can cross-present antigens and thus activate $CD8^+$ T cells as well (Collin and Bigley, 2018).

moDCs, as their name implies, arise from classical monocytes ($Ly6C^+$ and $CD14^+$ in mice and humans, respectively) instead of from common DC progenitors. They display high levels of MHC-II and CD11c and they also express Ly6C and CD64 in mice and CD64 and CD14 in humans (Gardner et al., 2020; Lee et al., 2018). They exert direct antimicrobial effector functions (Wculek et al., 2020), produce high amounts of IL-12 and seem to be able to cross-present antigens (Gardner et al., 2020; Murphy and Murphy, 2021).

3. Dendritic cells in the immune niche of the peritoneal cavity

The peritoneal cavity is the largest body cavity. It is filled with peritoneal fluid, which lubricates the surface of the abdominal viscera, and it is lined by a thin layer of mesothelial cells called peritoneum. These mesothelial cells produce many substances that regulate the activity of the peritoneal hematopoietic cells, and molecules that act as surfactants for the surface of viscera (van Baal et al., 2017). A significant number of immune cells are maintaining the tissue homeostasis and protecting the body from peritoneal infections and injury. Macrophages are the main immune cell population (Bou Ghosn et al., 2010), representing approximately half of the peritoneal leukocytes, and they are specialized in the clearance of pathogens and apoptotic cells (Roberts et al., 2017). They are also implicated in the maintenance of B lymphocytes through secretion of CXCL13 (Ansel et al., 2002).

Different DC subsets are also present in the peritoneal immune cell repertoire. A recent study carried out with isolated peritoneal exudate cells indicated that conventional peritoneal DCs are composed of 20% cDC1s and 80% cDC2s (Sohn et al., 2021). As mentioned before, the cDC1 subset includes $CD11c^+MHCII^+XCR1^+CD103^+$ cells, which are commonly described as Th1 inducer cells (Hildner et al., 2008) whereas the cDC2 subset comprises $CD11c^+MHCII^+CD11b^+$ cells.

Sometimes the criteria to distinguish peritoneal macrophages and DCs are not clear and, in mice, some authors described five subsets of myeloid mononuclear cells, including DCs and macrophages (Sohn et al., 2019). More specifically, these subsets are large peritoneal macrophages (LPMs), small peritoneal macrophages (SPMs), DCs, and two subpopulations of $MHCII^+CD11c^+CD115^+$ cells (one expressing $CD14^+CD206^+$ and the other being negative for these markers). According to these authors, DCs are distinguished as $CD11c^+$ $CD115^-$ cells and are divided into a minor subset of $XCR1^+CD24^+$ cells (cDC1 subset) and a major subset of $XCR1^-$ $CD24^-$ cells (cDC2 subset). Among the five main subsets described, only DCs and $MHCII^+CD11c^+CD115^+CD14^-CD206^-$ cells were demonstrated to be competent APCs but DCs were clearly superior in activating T cells and expanding cytokine producing cells. The other population had a phenotype similar to cDC2, they were able to generate partially activated T cells and had a greater ability to induce Tregs.

According to another recent work published by Sohn et al. (2021), isolated peritoneal cDC1s had a greater capacity in cross-presenting and in activating $CD8^+$ T cells compared to cDC2s. Moreover, cDC1s were superior in controlling the global gene expression of $CD8^+$ T cells and IL-2 was identified as the main regulating cytokine. The augmentation of IL-2 levels by cDC1s lead to an induction of activation markers such as CD25 and CD69 in $CD8^+$ T cells.

Furthermore, it was observed that both DC subsets are equally capable of activating $CD4^+$ T cells. Interestingly, these authors demonstrated that after LPS injection, cDC1s expressed higher levels of CD86 than cDC2s but other co-stimulatory molecules such as CD80 and CD40 were expressed more highly on cDC2s than on cDC1s. Moreover, cDC1s induced high expression of lung-homing markers on the activated $CD8^+$ T cells, such as CCR4, while cDC2s favored the expression of gut-homing molecules such as CCR9.

Other types of immune innate cells detected in the peritoneal cavity include neutrophils, monocytes, innate lymphoid cells-2 (ILC-2), and natural killer T (NKT) cells (Meza-Perez and Randall, 2017a; Okabe, 2021). ILC-2 cells were initially described as natural helpers of nuocytes (Neill et al., 2010) and play a key role in differentiation of B1 cells due to their secretion of IL-5. B1 cells, a subset of B lymphocytes that produces natural antibodies, are the main adaptive immune population present in the peritoneal cavity (Okabe, 2021). However, $CD4^+$ and $CD8^+$ lymphocytes are also present as well as Tregs, which are considered important for the regulation of the immune response.

4. Dendritic cells in the Omentum

The omentum is an apron-like structure of adipose tissue located in front of abdominal organs that promotes a variety of immunological responses, including inflammation, tolerance, and wound healing among others (Di Nicola, 2019). In mice, the omentum is a thin strip of fat but in humans it is far larger and constitutes the major depot of abdominal fat (Wilkosz et al., 2005). In both humans and mice, the omentum contains aggregates of lymphocytes called milky spots (MS), which are similar to the secondary lymphoid organs follicles (Liu et al., 2021b; Meza-Perez and Randall, 2017b). These MS are situated under the mesothelial cell layer and are well connected with the peritoneal cavity through the stomata, which are fenestrations between the mesothelial cells (Mironov et al., 1979) (See Fig. 1).

Unlike the secondary lymphoid organs, MS lack afferent lymphatic vessels. Therefore, stomata permit to collect cells and antigens from the peritoneal cavity, allowing the trafficking of immune cells in physiological and pathological conditions (Carlow et al., 2009). Efferent lymphatic vessels do exist, they originate close to MS between adipocytes and drain downstream to the lymph nodes. The omentum is a highly vascularized tissue and MS are disposed around glomerulus-like structures of blood vessels (Rangel-Moreno et al., 2009). Some of these vessels have markers of high endothelial venules (HEV) such as peripheral lymph node addressin (PNAd) and mucosal addressin cell adhesion molecule 1 (MAdCAM-1) (Bain et al., 2014). In general, these addressin molecules permit the recruitment of mucosal-homing lymphocytes from the blood. Because of their localization, MS work as filters for the peritoneal fluid and are able to induce immune responses against any non-self-agent or against pathogens present in the peritoneal cavity.

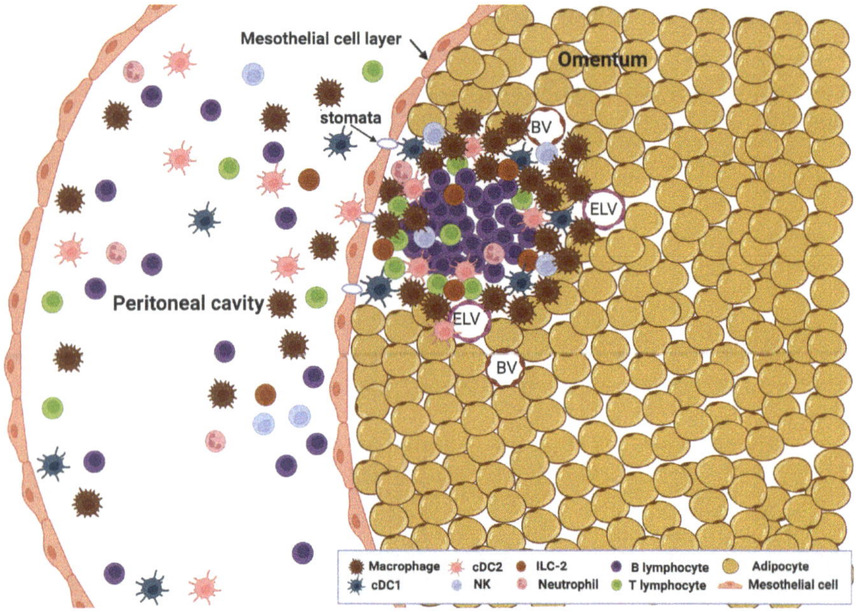

Fig. 1 Summary of the different immune cell populations present in the peritoneal cavity and omentum. BV: blood vessel; ELV: efferent lymphatic vessel. This figure was created using Biorender.

MS are composed of a central area of B cells (mostly B1 cells), which can migrate to peripheral lymph nodes and differentiate to IgM- and IgA-secreting cells. These B1 cells are surrounded by scarce T cells (Krist et al., 1997; Liu et al., 2021b) that are involved in the induction of cellular immune responses. There are also VAT (visceral adipose tissue)-associated Tregs that produce mainly IL-10 and are implicated in the regulation of metabolism in adipocytes (Cipolletta et al., 2015). Innate myeloid cells are also present in MS and macrophages are the major population with several versatile functions such as clearance of apoptotic cells, and phagocytosis of pathogens. When an inflammation process is initiated in the peritoneal cavity, a rapid migration of the macrophages to the omentum occurs, a phenomenon known as macrophage disappearance reaction (Meza-Perez and Randall, 2017b). LPMs predominate over the SPMs and usually express the transcription factor GATA6 as well as TGF-β (Okabe and Medzhitov, 2014). They also produce retinoic acid, an important factor for the maintenance of B1 cells. Moreover, IL-33 secreted by fibroblastic reticular cells from stroma and mesothelial cells is essential for macrophage polarization and induces anti-inflammatory M2 $CD206^{+}$ cells (Liu et al., 2021b).

Apart from the macrophage population, DCs are also present in the omentum. They promote rapid activation of T cells and cross-present not only soluble antigens but also antigens acquired by phagocytosis of pathogens and malignant cells. Like in the peritoneal cavity, in the omentum the cDC2 subset is the main DC population and makes up about 70% of the total DCs. Like macrophages, omental DCs express the enzyme retinaldehyde dehydrogenase-2, which is crucial for the production of acid retinoic, which is involved in the differentiation of mucosal homing T cells and is an important factor for the induction of VAT-associated Tregs. As described by previous studies, the activation of cDC2s in the peritoneal cavity is normally suppressed by the upregulation of the peroxisome proliferator-activated receptor γ (PPARγ) pathway (Macdougall et al., 2018). cDC1s can also be found in the omentum but despite being commonly described as IL-12 producing and Th1 inducers, in this environment they can produce IL-10 by upregulation of the Wnt/β-catenin pathway (Macdougall et al., 2018). In these conditions, cDC1s can induce Tregs and control Th17 responses. Thus, both mechanisms promote an anti-inflammatory microenvironment, which maintains adipocyte homeostasis and regulation of immune responses.

5. Dendritic cells in the Mesentery and gut-associated lymphoid tissue (GALT)

The mesentery is a connective tissue membrane formed by a double layer of peritoneum that covers the intestines and suspends them within the peritoneal cavity from the abdominal wall preventing their collapse. It contains nerves, blood and lymph vessels, lymph nodes (Coffey and O'Leary, 2016) and reminiscent fat omental milky spot-like structures called fat-associated lymphoid clusters (FALCs), which have the same function as MS (Okabe, 2021).

Since the intestines are continuously exposed to antigens derived from both food and commensal microorganisms (microbiota), a strong immune regulation is pivotal for homeostasis not only to maintain oral tolerance against ingested food, but also to initiate immune responses against potential pathogens. This regulation mainly depends on intestinal DCs, which are predominantly localized both in GALT (such as Peyer patches) and mesenteric lymph nodes (MLN; the largest lymph nodes of the body), the most important lymphocyte priming locations for immune responses in the intestine. In addition, intestinal DCs can also be found in lamina propia and scattered in lymphoid aggregates, called solitary isolated lymphoid tissues (Sun et al., 2020; Worbs et al., 2017).

In the human and murine mesentery, three subsets of cDCs have been described: cDC1s, CD103^{+} cDC2 (migratory DCs) and CD103^{-} cDC2 (resident DCs in MLN). Both human and murine cDC1 cells can express CD103 as well. Since macrophages also exhibit CD11b but lack CD103, markers such as CD64, F4/80 and CX3CR1 can be useful for macrophage identification and to distinguish them from cDC2s (Sun et al., 2020).

The different DC subsets have specific anatomical distributions. Within the Peyer's patches, cDC2s are located between the follicle and the epithelium that is covering the patch, in a zone called subepithelial dome, which is enriched in macrophages, B- and T lymphocytes. cDC1s in contrast, are localized in the T interfollicular region, enriched in T cells. Outside the Peyer's patches, the solitary isolated lymphoid tissues are mainly composed of cDC1s and pDCs (Sun et al., 2020). In contrast, in the lamina propia, MLN structures and in other gut-draining lymph nodes, cDC2s are generally the predominant DC population, except in the lamina propia of the colon region (Sun et al., 2020). After activation, DCs from intestinal lamina propia and Peyer patches migrate to the MLN via lymph vessels using the CCR7/CCL21 signaling axis (Jang et al., 2006; Worbs et al., 2017).

cDC1s express CXCL13, which recruits B cells (via CXCR5) to form B cell follicles scattered along the intestine (McDonald et al., 2010). After intestinal epithelial cell conditioning with TGF-β, thymic stromal lymphopoietin (TSLP) and retinoic acid, lamina propia-derived CD103^{+} cDCs acquire a tolerogenic phenotype that can induce Tregs (Iliev et al., 2009; Laffont et al., 2010; Rimoldi et al., 2005). These DCs are also involved in the recruitment of Tregs from MLN to the intestine by CCR9 upregulation (Sun et al., 2020; Worbs et al., 2017). These Treg induction and recruitment mechanisms are impaired during inflammation (Laffont et al., 2010). In such circumstances, more DCs can be generated from gut-resident and blood monocytes, giving rise to moDCs (Worbs et al., 2017), which, under physiological conditions, would differentiate to anti-inflammatory intestinal macrophages (CD64^{+} F4/80^{+} CX3CR1^{+}) (Bain et al., 2014). In addition to the generation of moDCs, DCs can be directly recruited from blood to the MLN, which can be highly beneficial in terms of T cell activation since they have not been previously conditioned by the tolerogenic environment of the gut (Laffont et al., 2010).

pDCs can be found in gut but not in MLN (Wendland et al., 2007; Worbs et al., 2017). They are recruited to the gut in a CCR9- dependent manner, guided by CCL25 produced by epithelial cells of the small intestine. Their role in intestinal immune reactions seems to be related to an increase on DC mobilization, presumably due to TNF-α secretion (Wendland et al., 2007). GALT pDCs along with cDCs in steady-state play an important role in antibody isotype polarization towards IgAs (Sun et al., 2020; Tezuka et al., 2011).

6. Concluding remarks

The peritoneal cavity is a complex compartment, which requires careful balancing of immunity versus tolerance. It contains a variety of immune cells that are continuously changing according to the physiological and pathological conditions, serving as a first defense line against pathogens, tissue damage and cancer metastasis in the peritoneum. In this scenario, DCs play a key role in orchestrating the balance between protective immune responses and the maintenance of metabolic and adipocyte homeostasis. While currently little is known about the precise mechanisms, they include the conditioning of DCs by the surrounding tissue and their promotion of an anti-inflammatory environment and the recruitment of Tregs. Future studies will help to further unravel the interactions between DCs and other (immune) cell populations and to elucidate the mechanisms that govern this unique immunological environment.

References

Ansel, K.M., Harris, R.B.S., Cyster, J.G., 2002. CXCL13 is required for B1 cell homing, natural antibody production, and body cavity immunity. Immunity 16 (1), 67–76.

Bain, C.C., Bravo-Blas, A., Scott, C.L., Gomez Perdiguero, E., Geissmann, F., Henri, S., et al., 2014. Constant replenishment from circulating monocytes maintains the macrophage pool in the intestine of adult mice. Nat. Immunol. 15 (10), 929–937.

Banchereau, J., Briere, F., Caux, C., Davoust, J., Lebecque, S., Liu, Y., et al., 2000. Immunobiology of dendritic cells. Annu. Rev. Immunol. [Internet]. 18 (Figure 1), 767–811. Available from: http://www.annualreviews.org/doi/10.1146/annurev.immunol.18.1.767.

Bou Ghosn, E.E., Cassado, A.A., Govoni, G.R., Fukuhara, T., Yang, Y., Monack, D.M., et al., 2010. Two physically, functionally, and developmentally distinct peritoneal macrophage subsets. Proc. Natl. Acad. Sci. U. S. A. 107 (6), 2568–2573.

Carlow, D.A., Gold, M.R., Ziltener, H.J., 2009. Lymphocytes in the peritoneum home to the omentum and are activated by resident dendritic cells. J. Immunol. 183 (2), 1155–1165.

Chen, B., Zhu, L., Yang, S., Su, W., 2021. Unraveling the heterogeneity and ontogeny of dendritic cells using single-cell RNA sequencing. Front. Immunol. 12, 1–13.

Cipolletta, D., Cohen, P., Spiegelman, B.M., Benoist, C., Mathis, D., 2015. Appearance and disappearance of the mRNA signature characteristic of Treg cells in visceral adipose tissue: age, diet, and PPARγ effects. Proc. Natl. Acad. Sci. U. S. A. 112 (2), 482–487.

Coffey, J.C., O'Leary, D.P., 2016. The mesentery: structure, function, and role in disease. Lancet Gastroenterol Hepatol [Internet]. 1 (3), 238–247. Available from: https://doi.org/10.1016/S2468-1253(16)30026-7.
Collin, M., Bigley, V., 2018. Human dendritic cell subsets: an update. Immunology 154 (1), 3–20.
Di Nicola, V., 2019. Omentum a powerful biological source in regenerative surgery. In: Regenerative Therapy. Japanese Society of Regenerative Medicine. Vol. 11, pp. 182–191.
Fucikova, J., Palova-Jelinkova, L., Bartunkova, J., Spisek, R., 2019. Induction of tolerance and immunity by dendritic cells: mechanisms and clinical applications. Front. Immunol. 10, 1–17.
Gardner, A., de Mingo, P.Á., Ruffell, B., 2020. Dendritic cells and their role in immunotherapy. Front. Immunol. 11, 1–14.
Guilliams, M., Ginhoux, F., Jakubzick, C., Naik, S.H., Onai, N., Schraml, B.U., et al., 2014. Dendritic cells, monocytes and macrophages: a unified nomenclature based on ontogeny. Nat. Rev. Immunol. 14 (8), 571–578.
Guilliams, M., Dutertre, C.A., Scott, C.L., McGovern, N., Sichien, D., Chakarov, S., et al., 2016. Unsupervised high-dimensional analysis aligns dendritic cells across tissues and species. Immunity 45 (3), 669–684.
Hasegawa, H., Matsumoto, T., 2018. Mechanisms of tolerance induction by dendritic cells in vivo. Front. Immunol. 9.
Hildner, K., Edelson, B.T., Purtha, W.E., Diamond, M., Matsushita, H., Kohyama, M., et al., 2008. Batf3 deficiency reveals a critical role for CD8α+ dendritic cells in cytotoxic T cell immunity. Science (80-) 322 (5904), 1097–1100.
Iliev, I.D., Spadoni, I., Mileti, E., Matteoli, G., Sonzogni, A., Sampietro, G.M., et al., 2009. Human intestinal epithelial cells promote the differentiation of tolerogenic dendritic cells. Gut 58 (11), 1481–1489.
Jang, M.H., Sougawa, N., Tanaka, T., Hirata, T., Hiroi, T., Tohya, K., et al., 2006. CCR7 is critically important for migration of dendritic cells in intestinal lamina propria to mesenteric lymph nodes. J. Immunol. 176 (2), 803–810.
Kambayashi, T., Laufer, T.M., 2014. Atypical MHC class II-expressing antigen-presenting cells: can anything replace a dendritic cell? Nat. Rev. Immunol. 14 (11), 719–730.
Krist, L.F.G., Koenen, H., Calame, W., Van Der Harten, J.J., Van Der Linden, J.C., Eestermans, I.L., et al., 1997. Ontogeny of Milky Spots in the Human Greater Omentum: An Immunochemical Study. Vol. 249 Wiley-Liss, Inc.
Laffont, S., Siddiqui, K.R.R., Powrie, F., 2010. Intestinal inflammation abrogates the tolerogenic properties of MLN CD103+ dendritic cells. Eur. J. Immunol. 40 (7), 1877–1883.
Lee, C., Lee, M., Rhee, I., 2018. Distinct features of dendritic cell-based immunotherapy as cancer vaccines. Clin Exp Vaccine Res [Internet]. 7 (1), 16. Available from: https://doi.org/10.7774/cevr.2018.7.1.16.
Liu, J., Zhang, X., Cheng, Y., Cao, X., 2021a. Dendritic cell migration in inflammation and immunity. Cell. Mol. Immunol. 18 (11), 2461–2471.
Liu, M., Silva-Sanchez, A., Randall, T.D., Meza-Perez, S., 2021b. Specialized immune responses in the peritoneal cavity and omentum. J. Leukoc. Biol. 109 (4), 717–729.
Macdougall, C.E., Wood, E.G., Loschko, J., Scagliotti, V., Cassidy, F.C., Robinson, M.E., et al., 2018. Visceral adipose tissue immune homeostasis is regulated by the crosstalk between adipocytes and dendritic cell subsets. Cell Metab. 27 (3), 588–601.e4.
Macri, C., Pang, E.S., Patton, T., O'Keeffe, M., 2018. Dendritic cell subsets. Semin Cell Dev Biol [Internet] 84, 11–21. Available from: https://doi.org/10.1016/j.semcdb.2017.12.009.
Mccully, M.L., Madrenas, J., 2006. Dendritic cells as arbiters of peritoneal immune responses. Perit. Dial. Int. 26, 8–25.

McDonald, K.G., McDonough, J.S., Dieckgraefe, B.K., Newberry, R.D., 2010. Dendritic cells produce CXCL13 and participate in the development of murine small intestine lymphoid tissues. Am. J. Pathol. 176 (5), 2367–2377.
Meza-Perez, S., Randall, T.D., 2017a. Immunological functions of the omentum. Trends Immunol. 38 (7), 526–536.
Meza-Perez, S., Randall, T.D., 2017b. Immunological functions of the Omentum. Trends Immunol. 38 (7), 526–536.
Mironov, V.A., Gusev, S.A., Baradi, A.F., 1979. Cell and tissue research mesothelial stomata overlying omental milky spots: scanning electron microscopic study. Cell Tissue Res. 201.
Murphy, T.L., Murphy, K.M., 2021. Dendritic cells in cancer immunology. Cell Mol. Immunol. [Internet], 1–11. Available from: https://doi.org/10.1038/s41423-021-00741-5.
Neill, D.R., Wong, S.H., Bellosi, A., Flynn, R.J., Daly, M., Langford, T.K.A., et al., 2010. Nuocytes represent a new innate effector leukocyte that mediates type-2 immunity. Nature 464 (7293), 1367–1370.
Okabe, Y., 2021. Immune Niche within the Peritoneal Cavity. Springer, Cham, pp. 123–134. [cited 2022 Jan 9]. Available from: https://link.springer.com/chapter/10.1007/978- 3-030-86016-5_6.
Okabe, Y., Medzhitov, R., 2014. Tissue-specific signals control reversible program of localization and functional polarization of macrophages. Cell 157 (4), 832–844.
Rangel-Moreno, J., Moyron-Quiroz, J.E., Carragher, D.M., Kusser, K., Hartson, L., Moquin, A., et al., 2009. Omental milky spots develop in the absence of lymphoid tissue-inducer cells and support B and T cell responses to peritoneal antigens. Immunity 30 (5), 731–743.
Reis, E., Sousa, C., 2006. Dendritic cells in a mature age. Nat. Rev. Immunol. 6 (6), 476–483.
Rimoldi, M., Chieppa, M., Salucci, V., Avogadri, F., Sonzogni, A., Sampietro, G.M., et al., 2005. Intestinal immune homeostasis is regulated by the crosstalk between epithelial cells and dendritic cells. Nat. Immunol. 6 (5), 507–514.
Roberts, A.W., Lee, B.L., Deguine, J., John, S., Shlomchik, M.J., Barton, G.M., 2017. Tissue-resident macrophages are locally programmed for silent clearance of apoptotic cells. Immunity 47 (5), 913–927.e6.
Smith, C.M., Wilson, N.S., Waithman, J., Villadangos, J.A., Carbone, F.R., Heath, W.R., et al., 2004. Cognate CD4 + T cell licensing of dendritic cells in CD8 + T cell immunity. Nat. Immunol. 5 (11), 1143–1148.
Sohn, M., Na, H.Y., Ryu, S.H., Choi, W., In, H., Shin, H.S., et al., 2019. Two distinct subsets are identified from the peritoneal myeloid mononuclear cells expressing both CD11C and CD115. Immune. Netw. 19 (3).
Sohn, M., Na, H.Y., Shin, H.S., Ryu, S.H., Park, S., In, H., et al., 2021. Global gene expression of T cells is differentially regulated by peritoneal dendritic cell subsets in an IL-2 dependent manner. Front. Immunol. 12, 1–16.
Steinman, R.M., Cohn, Z.A., 1973. Identification of a novel cell type in peripheral lymphoid organs of mice: I. morphology, quantitation, tissue distribution. J. Exp. Med. 137 (5), 1142–1162.
Sun, T., Nguyen, A., Gommerman, J.L., 2020. Dendritic cell subsets in intestinal immunity and inflammation. J Immunol [Internet]. 204 (5), 1075–1083. Available from: http://www.jimmunol.org/content/204/5/1075.
Tezuka, H., Abe, Y., Asano, J., Sato, T., Liu, J., Iwata, M., et al., 2011. Prominent role for plasmacytoid dendritic cells in mucosal T cell-independent IgA induction. Immunity [Internet]. 34 (2), 247–257. Available from: https://doi.org/10.1016/j.immuni.2011.02.002.
Thaiss, C.A., Semmling, V., Franken, L., Wagner, H., Kurts, C., 2011. Chemokines: a new dendritic cell signal fort cell activation. Front. Immunol. 2, 1–9.

van Baal, J.O.A.M., Van de Vijver, K.K., Nieuwland, R., van Noorden, C.J.F., van Driel, W.J., Sturk, A., et al., 2017. The histophysiology and pathophysiology of the peritoneum. Tissue Cell [Internet]. 49 (1), 95–105. Available from: https://doi.org/10.1016/j.tice.2016.11.004.

Villar, J., Segura, E., 2020. Decoding the heterogeneity of human dendritic cell subsets. Trends Immunol [Internet] 41 (12), 1062–1071. Available from: https://doi.org/10.1016/j.it.2020.10.002.

Wculek, S.K., Cueto, F.J., Mujal, A.M., Melero, I., Krummel, M.F., Sancho, D., 2020. Dendritic cells in cancer immunology and immunotherapy. Nat. Rev. Immunol. [Internet]. 20 (1), 7–24. Available from: https://doi.org/10.1038/s41577-019-0210-z.

Wendland, M., Czeloth, N., Mach, N., Malissen, B., Kremmer, E., Pabst, O., et al., 2007. CCR9 is a homing receptor for plasmacytoid dendritic cells to the small intestine. Proc. Natl. Acad. Sci. U. S. A. 104 (15), 6347–6352.

Wilkosz, S., Grenham, A.E., Nadeem, I., Ae, K., Ae, M.W., Butt, R., et al., 2005. A comparative study of the structure of human and murine greater omentum. Anat. Embryol. 209, 251–261.

Worbs, T., Hammerschmidt, S.I., Förster, R., 2017. Dendritic cell migration in health and disease. Nat. Rev. Immunol. 17 (1), 30–48.

CHAPTER TWO

T cells in the peritoneum

Sanjay Rathod[†,*]
Department of Immunology, University of Pittsburgh, Pittsburgh, PA, United States
*Corresponding author: e-mail address: sanjayrathod2dec@gmail.com

Contents

Abstract

The peritoneal cavity is a fluid-packed area that houses most of the abdominal organs, including the omentum, a visceral adipose tissue with milky patches or groups of leukocytes organized in the same way to those observed in typical lymphoid tissues. A distinct population of leukocytes patrols the peritoneal cavity and travels in and

[†] *Current address*: Novasenta Inc., Pittsburgh, PA 15219, United States.

International Review of Cell and Molecular Biology, Volume 371
ISSN 1937-6448
https://doi.org/10.1016/bs.ircmb.2022.04.013

out of the milky spots, facing antigens or pathogens in the peritoneal fluid and responding appropriately. T cells may play a crucial function in regulating adaptive immune responses to antigens in the peritoneal cavity to ensure tissue homeostasis and healing. When peritoneal homeostasis is interrupted by inflammation, infection, obesity, or tumor metastasis, the omentum's dedicated fibroblastic stromal cells and mesothelial cells control peritoneal leukocyte recruitment and activation in unique ways. T cells, which employ their T cell receptor to target specific antigens, are an important component of the acquired immune response since they are present in the peritoneal cavity. The peritoneum provides a different environment for T cells to respond to pathogens. This chapter outlines the anatomy relevant to T cell function and biology, such as antigen processing/presentation, T cell activation, and the many T cell subpopulations in the peritoneal cavity, as well as their role in cancer or other infection.

1. Introduction

The name peritoneum is derived from the Greek words peri (around) and tonos (within or "stretched"). The peritoneum is a thick stroma layer that surrounds the innermost surface of the abdominal wall as well as several of the abdominal organs. A single sheet of mesothelial cells covers its inner surface. The peritoneum is confined in men, whereas it is only accessible to the outside at the Ostia of the fallopian tubes in women. The peritoneum is separated into two sections: the parietal and visceral peritoneum. The parietal peritoneum covers the innermost surface of the abdominal walls, the inferior surface of the diaphragm, and the pelvis. The visceral peritoneum, on the other hand, covers the bulk of the intraperitoneal organs as well as the front portion (but not the entire) of the retroperitoneal organs (reviewed in Meyers et al., 1987). Despite its intricacy, the peritoneum is frequently misunderstood in classical medical texts and is merely referred to as the surrounding tissue (serous membrane) of the gut. There are discoveries in the physiology and morphology of the peritoneum and mesothelial cells. According to recent study on the anatomy, histology, and physiology of the peritoneum, it is a more complex structure than a simple serous membrane.

A serosal membrane that lines the abdominal cavity is known as the peritoneum. The peritoneum is the first line of defense against pathogens and malignant cells in the body. The peritoneal cavity is a specialized compartment with a fluid-filled region that houses the majority of the abdominal organs, including the omentum, a visceral adipose tissue with milky patches or clusters of immune cells organized similarly to those found in conventional lymphoid tissues. The peritoneum contains the bulk of immune cells, which include macrophages and B1 cells, with fewer T cells, eosinophils, neutrophils, natural killer cells (NK), and dendritic cells (DCs) (DCs) (Ghosn et al., 2010; Terri et al., 2021).

A separate population of leukocytes patrols the peritoneal cavity and migrates in and out of the milky patches, encountering Ags or pathogens in the peritoneal fluid and responding appropriately. Leukocytes' principal purpose in the peritoneal cavity is to maintain tissue homeostasis and to promote tissue healing. When peritoneal homeostasis is interrupted by inflammation, infection, obesity, or tumor metastasis, the omentum's specialized fibroblastic stromal cells and mesothelial cells regulate peritoneal leukocyte recruitment and activation in unique ways. This chapter will examine the types of cells found in the peritoneal cavity depicted in Fig. 1, the role of the omentum in their maintenance and activation, and how these mechanisms occur in response to infections and cancer.

The peritoneal cavity is a distinct compartment in which infections can quickly spread to neighboring organs such as the gut and liver. The peritoneal cavity connects with the systemic blood circulation via the omental milky spot, a lymphoid tissue embedded in omental fat (Rangel-Moreno et al., 2009). The largest body cavity is the peritoneal cavity, which is enclosed by mesothelial layers and houses the stomach, spleen, intestines, pancreas, reproductive organs, and so on. The peritoneum is a bidirectional, semipermanent biomembrane that controls the amount of fluid in the peritoneal cavity. The peritoneal cavity typically contains less than 100 mL of sterile peritoneal fluid in a healthy individual that lubricates the surface of visceral tissues (Healy and Reznek, 1998). The volume of peritoneal fluid increases in pathological circumstances such as liver cirrhosis, cancer, and pancreatitis. Ascites is caused by an increase in capillary pressure in the abdominal viscera, which permits fluid to enter the peritoneal cavity (Szender et al., 2017). As a result, the peritoneal environment fluctuates dynamically in response to a variety of physiological and pathological conditions.

The peritoneal membrane in healthy persons was composed of a monolayer of mesothelial cells on a basement membrane and a layer of connective tissue that included cells, blood vessels, and lymphatic vessels. Mesothelial cells were polygonal and frequently elongated, with a dense network of microvilli on their luminal surface. The peritoneal membrane is the smooth, translucent membrane that lines the abdominal cavity and includes the internal organs of the abdomen and pelvis, such as the stomach and large intestine.

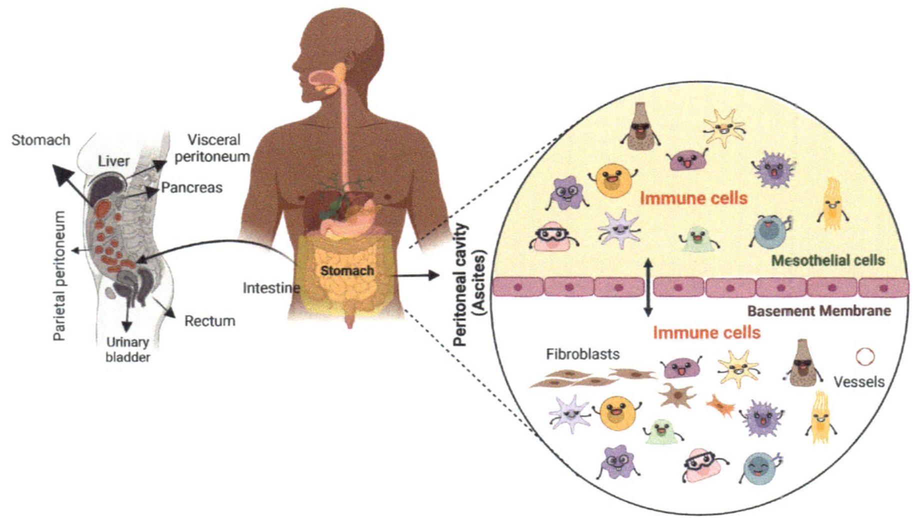

Fig. 1 Peritoneal anatomy and immune niche.

The peritoneal membrane is a membrane that protects and isolates the internal abdominal and pelvic tissues.

Many immune cells, including macrophages, mast cells, T cells, and B cells, are found in the peritoneal cavity. Furthermore, the peritoneal cavity is a resource-rich habitat for the survival of B1 cells, which release natural antibodies (Abs) constitutively (Hardy, 2006). Peritoneal B2 cells have been shown to have B1 cell-like properties when compared to splenic or lymph node B2 cells (Hastings et al., 2006), while peritoneal T cells have received less attention. Furthermore, there are just a few cases of $CD4^+$ T cells assisting B1 cells (Erickson et al., 2001). Because the peritoneum is densely packed with blood vessels, it is commonly used as a systemic injection route for small experimental animals. Peritoneal DCs must be responsible for capturing, processing, and presenting Ags to T cells following an intraperitoneal (i.p.) injection, hence initiating immunological responses; nevertheless, research on those properties of peritoneal T cells has been limited. Even though recent research has revealed the diversity of peritoneal T lymphocytes and their subsets.

1.1 Lymphoid cells associated with peritoneum

Lymphoid cells are created from lymphoid progenitor stem cells; they provide lifelong immunity against infection and play an essential role in autoimmunity and cancer therapies. The size divides lymphoid cells into small (T and B) and large granular (NK cells). Lymphocytes are also called white blood cells ($CD45^+$ leukocytes) that provide an immune response against the autoimmune cell or foreign substances like pathogenic antigens. There are several key classes of lymphocytes; one of them is B cells, which are generated and are maturated in the bone marrow, whereas T lymphocytes are generated in the bone marrow and mature in the thymus. Lymphocytes are adaptive immune system cells after creation and maturation and produce antigen-specific immune responses to malignancies by recognizing the presence of tumor antigen in cancer cells or cells with antigen (APCs). These lymphocytes effectively respond to antigen or foreign protein encounters, differentiating themselves into the active form as effector phenotype and the reservation form as memory cells. The active T effector cells can make cytotoxic granules or produce unique signals for other immune cells to achieve the adaptive immune response. A few activated effector T cells do not experience cell death, leading to memory T cells phenotype that persists in the peripheral lymphoid organs and remains potent to antigen recalls. In contrast, B cells encounter an antigen; they differentiate into antibody-producing plasma cells and memory cells. Memory cells stay longer and produce additional antibodies for future encounters with the same antigen. Thus, NK cells fit the subtype of lymphoid cells dealing primarily cytotoxic responses.

1.2 T cell and its subsets

T cells play a crucial role in the eradication of pathogens and the scrutiny of pathological cells. T cells are of two major subsets founded on their unique T cell receptor (TCR) chain, cytokine expression profile, and co-receptor expression. Based on TCR, they can discriminate between self-and non-self-antigens. During T cell development in the thymus, its precursors obligate into two T cell lineages based on expressing different TCR chains, either αβ T cells or γδ T cell (Fig. 2A). According to their co-receptor expression, most human T cells are αβ T cells (∼95%), further subdividing into CD4 T cells and CD8 cytotoxic T lymphocytes (CTLs). These effector

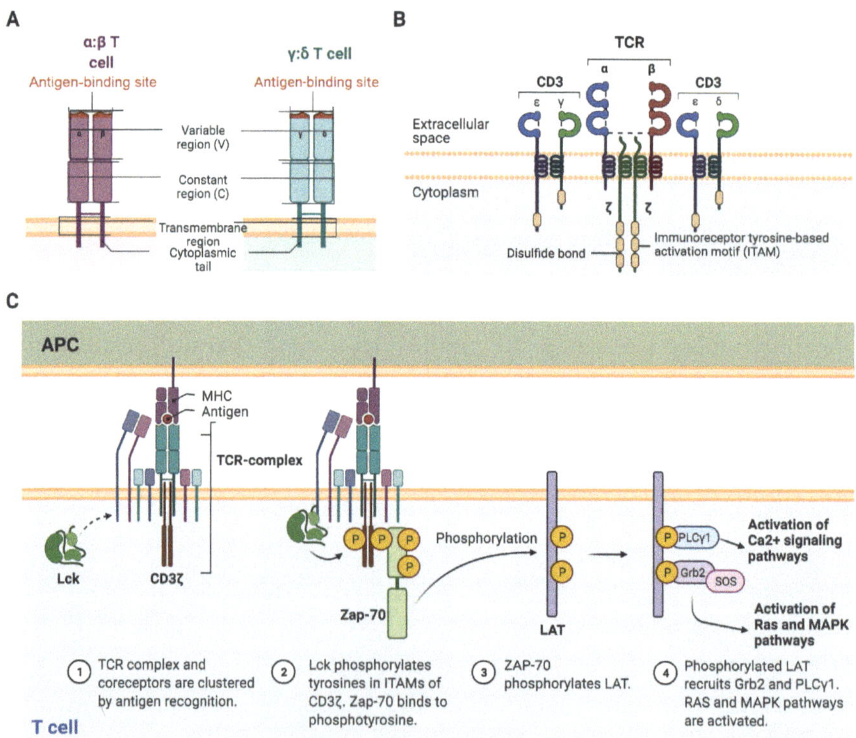

Fig. 2 TCR types, TCR subunits, and TCR signaling.

CD8 and CD4 T cells activation happen upon cognate TCR ligation to antigens presented on MHC-I and MHC-II molecules by APCs. The TCR-CD3 complex is made up of a variety of TCR heterodimers that are noncovalently linked to the invariant CD3 dimers CD3εγ, CD3εδ, and CD3ζζ (Kuhns and Badgandi, 2012; Wucherpfennig et al., 2010). TCRs mediate recognition of antigenic peptides attached to MHC molecules (pMHC), whereas CD3 molecules provide activation signals to T cells. The downstream T-cell signaling pathways, the mechanism by which TCR engagement by pMHC is first conveyed to the CD3 signaling apparatus, a process known as early T-cell activation summarize in Fig. 2B, C (detailed reviewed in Mariuzza et al., 2020).

Depending on the cytokine environment, CD4 T cells further differentiate into Th1, Th2, Th17, or Treg subsets (Fig. 3), which have different functions and cytokine expression profiles (Riera-Domingo et al., 2020). While other helper T cell subsets (such as Th3, Th9, and Th22) have recently been discovered, their precise function in peritoneum and cancer remains unknown and is not covered in this chapter.

Immune cells, especially T cells, are vital players in the immune system to fight against virus/cancer because they mount antigen-specific responses against viral infected or tumor cells. A few tumor-specific activated T cells directly destroy cancer cells by various mechanisms: (i) secreting cytotoxic molecules; (ii) releasing signaling molecules, such as cytokines, that prime and activate other types of immune cells (Fig. 3). The metabolic switches are crucial in the activation and differentiation of T cells from emerging evidence. The metabolism-based approaches for treating cancer appear to be the best strategy, and identifying new metabolite targets may help immuno-based therapy. This section emphasizes recent advances in metabolic switches that affect the T cells and their subtype proliferation,

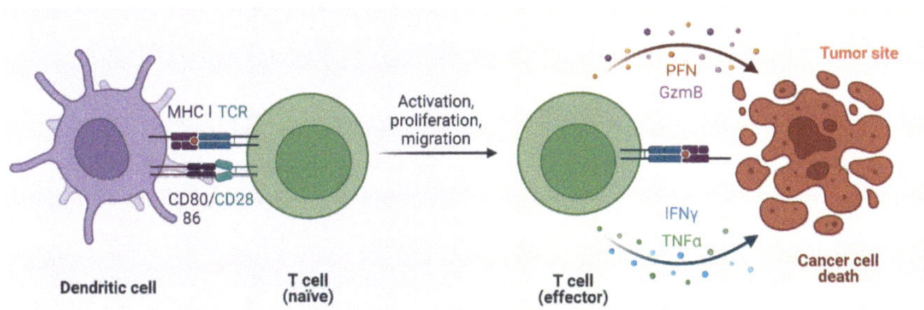

Fig. 3 Activation, proliferation, and migration of T cells.

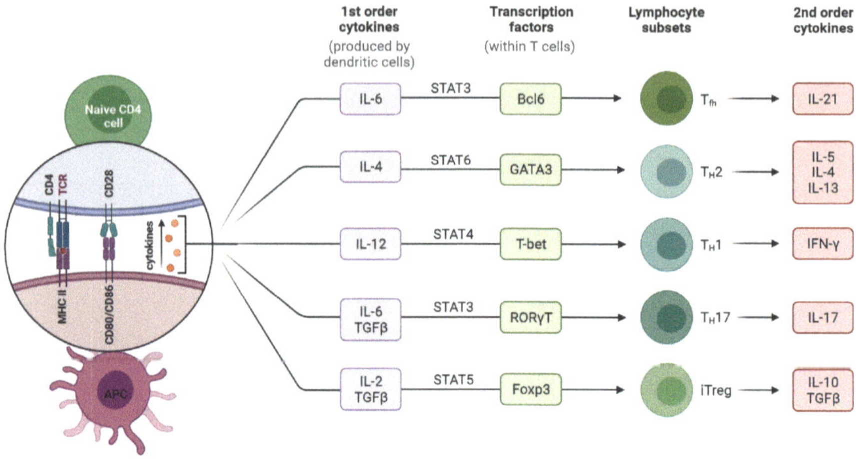

Fig. 4 T cell activation and differentiation.

differentiation, and execution of effector functions. T cell activation, expansion, and ultimately, differentiation into different subsets cytotoxic, Tregs and helper T cells (Th1, Th2, Treg, and Th17) after engaging with the first signal through T cell receptor (TCR) and second as co-stimulatory signals summarized in Fig. 4 (Swain et al., 2012; Wang et al., 2011). On cancer cells, activated cytotoxic $CD8^+$ T cells use a specific and effective cytotoxic activity.

Activated helper $CD4^+$ T cells, on the other hand, differentiate into effector phenotypes that either support or suppress tumor progression. The best way to convey this, the Th1 subset ($CD4^+IFN\gamma^+$) secrete the anti-cancer IFN-γ cytokine, which leads to stimulation of the NK and macrophages, and further initiate the anti-cancer immune response, while the Th2 subset ($CD4^+IL4^+$) and CD4 Treg subset ($CD4^+CD25^+FoxP3^+/CD127^-$) suppress the immunity against the tumor or pathogens. The other subsets of helper CD4 T cells, Th17 cells ($CD4^+IL17^+$), can either support or stop tumor growth (Bailey et al., 2014).

1.3 T cells in the peritoneum

T cell activation depends on external antigens processing and effective presentation by APCs in lymph nodes, as well as on presentation by DCs and other APCs in infected tissue and peritoneal fluid. As a result, we initially briefly discuss the peritoneum's T cell-relevant anatomy, with a focus on mesothelial cells and lymph nodes. The mesothelium is a layer of mesodermal cells that lines the peritoneal cavity. This layer of peritoneal mesothelial cells (PMCs) sits on top of a basement membrane, which in turn covers a 1–2 mm-deep submesothelial stroma of areolar tissue (Abu-Hijleh et al., 1995; Cameron, 1995). The peritoneum overlying the muscular component of the diaphragm has a slightly different structure. The PMC layer is disrupted in this section of the peritoneum by a high number of intercellular spaces (stomas) that lie between the lateral margins of the mesothelial cells (Abu-Hijleh et al., 1995). Because there is no basement membrane in this area, the stomas can interact with the underlying terminal lymphatic lacunas. These lacunae drain into the main lymphatic channels through the mediastinal lymph nodes. This phenomenon is reinforced by the discovery at autopsy that the anterior mediastinal lymphatics of peritonitis patients contain germs. The predominant components of the submesothelial stroma are glycosylated proteins, with macrophages, mast cells, and lymphocytes appearing on occasion (Heel and Hall, 1996).

PMCs as major regulators of intraperitoneal homeostasis, particularly as active participants in peritoneal immune defense. Because of their position between the peritoneal cavity and the peritoneal blood arteries, they play an important role in intraperitoneal immune protection. Mesothelial cells play an active role in the inflammatory response by generating inflammatory cytokines like IL-1α, IL-1β, IL-6 as well as chemokines such as IL-8, monocyte chemoattractant protein-1 (MCP-1), and RANTES (Douvdevani et al., 1994; Hurst et al., 2001; Jonjić et al., 1992; Li et al., 1998; Topley et al., 1993; Zeillemaker et al., 1995). The peritoneum contains approximately 100 mL of fluid that works as a lubricant; it also includes proteins and certain leukocytes that give to the peritoneum's defense mechanisms. In a healthy individual, 90% of monocytes/macrophages, 5% of lymphocytes, and 5% polymorphonuclear cells (PMNs) are in the peritoneal fluid (Ganguly et al., 1980; Goldstein et al., 1984).

This immune cell percentage dramatically changes during the diseases like peritonitis (Glik and Douvdevani, 2006). The peritoneal lymphatic system drains the peritoneal fluid, carrying pathogens and foreign antigens with it to be exposed to T and B lymphocytes in the lymphatic nodes.

The omentum, another essential immunological component peculiar to the peritoneum, localizes infection by sticking to sites of inflammation or perforation. Other than the diaphragmatic stomas, the omentum is the only place within the peritoneal cavity that can absorb foreign particles and microorganisms (Platell et al., 2000; Shipley and Cunningham, 1916). It contains cell aggregates known as "milky spots," after the eminent French anatomist Ranvier (Hall et al., 1998). PMNs, macrophages, and lymphocytes are found in these areas. These milky spots are composed of 10% T lymphocytes, 10% B lymphocytes, major portion, i.e., 70% macrophages (Beelen, 1991). During an intraperitoneal infection, milky patches appear as small, cotton wool-like lumps to the naked eye.

Peritonitis is an inflammatory process that involves all the components of the innate and adaptive arm of the immune system summarized in (Fig. 5). There are two fundamental types of immune responses. The first is the innate response, which uses the inborn immune arsenal to recognize pathogen-associated motif patterns to react rapidly to fight the invasion. The second is the acquired response, which develops after birth and improves upon repeated exposure to a given pathogen. In most cases, innate immune cells initiate the first defense against invading pathogens as shown in the upper left panel of Fig. 5. Polymorphonuclear cells (PMNs), macrophages (M), natural killer cells (NK), and T cells are the major cells engaged in innate immunity. T cells, unlike other T cells, can respond to antigens without being presented by antigen-presenting cells. Following activation, PMNs release the soluble form of the IL-6 receptor (sIL-6R), activated M expresses the CD40L (CD154), and T cells produce interferon-gamma

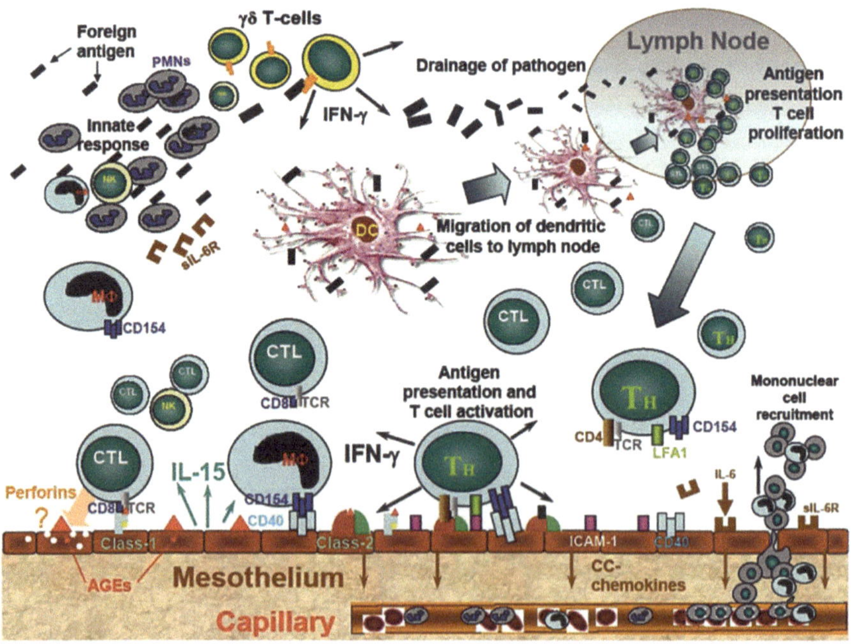

Fig. 5 T cells in the peritoneum.

(IFN-γ). These released factors are necessary for PMC activation. IFN-γ is particularly critical for the induction of class 2 molecules on PMCs. Pathogens in the peritoneal fluid are evacuated to the lymphatic system and collected in lymph nodes, where dendritic cells (DCs) offer their antigens to T cells (upper left). DCs may move from the peritoneum to the lymph node to present antigens that they have processed. T helper (Th) cells and cytotoxic T lymphocytes (CTLs) differentiate, proliferate, and move from lymph nodes to the peritoneum when stimulated. Antigen displayed on major histocompatibility complex (MHC) class 2 molecules expressed on PMCs may activate T lymphocytes. CD4 molecules are required to react with the MHC, the T cell receptor (TCR) identifies the given antigen, and leukocyte function-associated antigen (LFA)-1 binds intercellular adhesion molecule (ICAM)-1 on PMCs. Some activated T cells may also express CD154, which stimulates its receptor (CD40) on PMCs. IL-6 may activate PMCs that have recruited sIL-6R and, in conjunction with CD40 and IFN signals, produce a family of CC chemokines such as RANTES and monocyte chemoattractant protein-1, which stimulates endothelial cells to recruit mononuclear cells (lower right). IFN-γ produced by activated T cells and CD40 ligation stimulate the production of IL-15 by PMCs (lower left). IL-15 stimulates T cell proliferation and is required for the generation of NK cells and CD8+ memory T cells. High quantities of glucose breakdown products in peritoneal dialysis fluids are thought to produce an excess accumulation of advanced glycation end products (AGEs) in the peritoneum during CAPD treatment. Glycated cell membrane proteins on PMCs are likely detected as foreign antigens by CTLs, causing these cells to release perforins, which induce death in the target cell (Fig. 5).

2. Effector T cells

Omental T cells from patients with peritonitis have a Th2-like phenotype and produce more IL-4 while producing less IFN and TNF-α than those from patients with localized intra-abdominal infection (Russ and Reith, 2003), indicating that Th2 polarization is associated with immunosuppression in peritonitis patients and contributes to their increased mortality. T cells and mucosal-associated invariant T cells are found in the peritoneal cavity of individuals with acute peritonitis (Liuzzi et al., 2016), implying that T cells in this area respond to non-conventional Ags. Effector T cells, particularly $CD8^+$ T cells, play a crucial role in tumor immune regulation. $CD8^+$ T cells from tumor-bearing omenta display higher levels of CD69 and lower levels of CD62L than $CD8^+$ T cells from blood (Lysaght et al., 2011), indicating that they are active but fail to inhibit tumor growth. Consistent with this discovery, tumor-reactive $CD8^+$ T lymphocytes enter the tumor-bearing omentum in mice but quickly exhaust (Flies et al., 2016).

HLA-DR, an activation marker, is significantly expressed by $CD4^+$ effector T cells (Teff) in the ovarian tumor-bearing omentum (Kovacsovics-Bankowski et al., 2014), but these cells have limited functional potential. Not surprisingly, Ms in tumor-bearing omenta expresses high levels of PD-L1 (Kovacsovics-Bankowski et al., 2014) but low levels of IL-12 (Saccani et al., 2006). PD-L1 inhibition, on the other hand, restores $CD4^+$ T cell recruitment to the omentum. T cell dysfunction in the omentum may potentially be related to hypoxic zones that promote tumor cell self-renewal while suppressing T cell effector functions (Jayaprakash et al., 2018; Miao et al., 2014; Scharping et al., 2017).

The majority of $CD8^+$ T cells in malignant ascites are Teff or effector memory (TEM) cells, and they display activation markers such as CD69 as well as a large number of inhibitory receptors (Bamias et al., 2007; Lieber et al., 2018; Nakano et al., 2018). Tumor-specific $CD8^+$ T lymphocytes exhibit high amounts of PD-1 and generate little IFN in mice (Flies et al., 2016). PD-L1 is abundantly expressed in ovarian cancer patients' ascites and inhibits $CD8^+$ T cell-mediated cytotoxicity (Abiko et al., 2013). $CD4^+$ TEM are numerous in ovarian cancer patients' ascites,188 where they express high amounts of the inhibitory receptors CTLA4 and PD-1 (Landskron et al., 2015), exhibit limited proliferation, and release little IFN-γ (Wefers et al., 2018).

3. T helper cells

Th lymphocytes are a vital cell population found in the peritoneum. Th cells are classified into four subpopulations; Th1, Th2, Th17, and Treg (Rathod, 2022). Th1 cells normally release IL-2 and IFN-γ, whereas Th2 lymphocytes mostly secrete IL-4, IL-5, IL-10, and IL-13, Th17 secreted by IL-17 and Treg cells secreted the IL-10 and TGF-β suppressive cytokines (Fig. 4). In general, Th1 cells promote cell-mediated immunity. Th2 cells are in charge of humoral immunity (Nakano et al., 2018). It is worth noting that the products of Th1 and Th2 cells inhibit the opposite cell type in a reciprocal manner (Abiko et al., 2013).

The microenvironment is most likely the most important aspect of this process. Interleukin-4 is required for Th2 cell induction (Swain et al., 1990) and is generated by memory $CD4^+$ cells. T cells, NK T cells, mast cells, and so on; IL-12 is required for Th1 cells. DCs and macrophages generate it (Seder et al., 1993) during induction. Dendritic cells, which are the primary APCs for T cells, also play an important role in T cell polarization. The way DCs work is necessary to trigger the formation of Th cell subsets. Depending on how DCs are activated Pathogens and parasites some elements resulting from the contaminated environment DCs are polarized by cells, which biases Th cell differentiation accordingly (Kapsenberg, 2003; Rissoan et al., 1999). Cell-mediated immunity is required for a successful response to bacterial infection. Indeed, peritonitis in CAPD patients is mostly a type 1 immune response, as seen by the early production of TNF-α, IL-1, IL-12, and IL-18, followed by an increase in IFN-γ (Brauner et al., 1996; Wang and Lin, 2005). The Th1 phenotype seen in peritoneal effluents appears to be linked to a decreased incidence of peritonitis (Chiesa et al., 2004; Lamperi and Carozzi, 1988).

4. Cytotoxic T lymphocyte (CD8+ CTLs)

High levels of glucose degradation products (GDPs) in peritoneal dialysis fluids are thought to generate an excess buildup of AGEs in the peritoneum during CAPD treatment, leading to functional and structural alterations in the peritoneal membrane of CAPD patients (Tauer et al., 2003). This event begs the question of whether CTLs recognize glycated cell membrane proteins as foreign antigens. This idea is supported by the continuous discovery of a greater percentage of activated T cells in lymphocytes isolated from CAPD patient effluents compared to peripheral lymphocytes (Davies et al., 1989; Lewis et al., 1993; Valle et al., 1989). These activated T cells comprise both CTLs and Th cells, and their presence shows that some level of local activation is present in the peritoneum of CAPD patients even when there is no inflammation (Lewis et al., 1993). T cell activation may be mediated via the AGE receptor (RAGE), as proposed by Schwenger et al., who found an increase in T cells in wild-type animals exposed to GDP but not in RAGE-deficient mice (Schwenger et al., 2006).

CTLs' primary functions are to remove virus-infected cells (Tripathy et al., 2012a) and to eradicate malignant cells that have been altered. A cell marks itself as a target for cytotoxic T lymphocytes by displaying non-self-peptides on its surface produced from its protein-coupled to MHC class 1. Glycosylated peptides, on the other hand, are provided on class 1 molecules and effectively trigger CTLs (Haurum et al., 1999). As a result, CTLs may be involved in the degradation of the peritoneal membrane following repeated exposure to peritoneal dialysis fluid. The effector CTL kills target cells in two ways. The first process includes inserting perforins into the target cell membrane, which allows the insertion of granzymes, which are unique enzymes that mediate apoptosis in the target cell. Fas ligand expressed on CTLs, on the other hand, can bind Fas molecules on target cells to cause apoptosis. Cytotoxic T lymphocytes can also release cytokines such as TNF-α, lymphotoxin, and IFN-γ, which aid in the resistance of neighboring cells to infection (Delves and Roitt, 2000). After the pathogen is neutralized, the expanded antigen-specific T cell pool contracts significantly due to apoptosis, and only 10% of the antigen-stimulated T cells survive as memory cells. Because of their increased precursor frequency compared to the naive repertoire, memory $CD8^+$ T cells provide greater protection following reinfection, as well as their quick capacity to multiply and carry out effector tasks at the site of infection (Becker et al., 2002). The mechanism by which CTLs get activation, growth, and survival cues from other cell types is unknown. The conventional hypothesis of T lymphocyte-DC cross-presentation postulates that DCs collect antigens from necrotic or apoptotic cells and deliver them to $CD8^+$ T cells via the MHC class 1 complex. $CD4^+$ T cells are likely required for establishing an efficient primary CTL response and producing memory $CD8^+$ T cells, but Th cells are not required for secondary expansion during the recall response (Shedlock and Shen, 2003). Alternatively, if APCs' Toll-like receptors (TLRs) are triggered by bacterial products, the need for Th cells for CTL stimulation may be avoided (Janeway Jr. and Medzhitov, 2002).

5. Mucosal-associated invariant T cells (MAIT)

Mucosal-associated invariant T (MAIT) cells are a type of T cell that acts as a bridge between the adaptive and innate arms of the immune system. They make about 1–10% of total circulating T cells in the blood and the bulk of innate-like lymphocytes in the human liver (Dusseaux et al., 2011; Godfrey et al., 2015). The presentation of microbial vitamin B metabolites coupled to the MHC class I-like protein MR1 on antigen-presenting cells is essential for the activation of their distinct T-cell receptor (TCR), which in humans has the invariant alpha chain Vα7.2Jα33. In the absence of antigen presentation by IL-12, these cells can be activated TCR-independently (Corbett et al., 2014; Jesteadt et al., 2018; Kjer-Nielsen et al., 2012).

MAIT cell number, phenotype, and function in the peritoneal cavity of individuals with different illnesses in the absence or presence of spontaneous bacterial peritonitis (SBP). Unlike cMAIT cells, which are depleted in end-stage liver disease and exhibit signs of functional exhaustion (Böttcher et al., 2018; Hegde et al., 2018; Riva et al., 2018), pMAIT cells from decompensated cirrhosis patients respond to bacterial products in a manner comparable to healthy MAIT cells and remain potent producers of TNFα and IFN-γ in an MR1-restricted fashion.

MAIT cells could relocate to sites of infection and release inflammatory cytokines and cytotoxic granules on activation, it plays important role in bacterial infection. Bacterial infection is the most common cause of death in people with advanced liver disease, especially when there is decompensated cirrhosis with ascites (Borzio et al., 2001; Caly and Strauss, 1993). The transfer of live bacteria from the stomach through the gastrointestinal barrier into mesenteric lymph nodes or the portal vein, followed by seeding into the ascitic fluid (AF) with limited antimicrobial capacity, causes spontaneous bacterial peritonitis (SBP) (Wiest et al., 2014). Immunologic control fails on numerous levels in patients with severe cirrhosis. Few investigations suggest both conventional and unconventional T cells have reduced adaptive immune responses in patients with advanced cirrhosis (Albillos et al., 2014; Peter et al., 2013; Riva et al., 2018).

6. γδ T cell population

According to various studies, γδT cells play a crucial role in immunological defense against infections and in regulating chronic inflammation (reviewed in Ziegler, 2004). T cells that express γδ TCRs are referred to as γδT T cells, instead of the αβ TCRs seen on conventional T cells shown in Fig. 2. The γδT cells account for only a tiny proportion around 1–5% of lymphocytes that circulate in the blood and peripheral organs, while conventional T cells are ~95%. γδT are common in epithelial-rich tissues such as the skin, gut, and reproductive tract, where they can account for up to 50% of the population. γδT cells are immune cells. Their populations have been seen to rise in the 24h after abdominal surgery, peritoneal fluid (Decker et al., 2005).

These cells identify antigens from bacteria and tumors, which are often tiny pyro phosphomonoesters and alkylamines and have an important part in the elimination of viral infection (Münz et al., 2005). It has been demonstrated, for example, that the Vaccinia virus causes a large increase in IFN-γ producing T cells. As early as the second post-infection day, bacteria were found in the peritoneal cavity. Vaccinia virus-infected mice were also lacking in γδT lymphocytes compared to control mice, have significantly greater Vaccinia virus titers, and have higher mortality (Selin et al., 2001). Most γδT cell subpopulations do not require antigen processing or MHC complex involvement. However, the Non-classical MHC class 1 molecule stimulates unique immune responses. γδT cells are immune cells. MICA and MICB are recognized by human intestinal T cells as well as a subpopulation of peripheral T cells. The non-classic MHC also limits the number of blood T cells. CD1c is a CD1 class 1 molecule that contains microbial glycolipids (Münz et al., 2005). T cells may serve as a connection between the innate and adaptive immune systems via controlling the activation of adaptive immune responses of NK cells, DCs, and other APCs via cytokine secretion TNF-α and IFN-γ are two examples. They may also help to guide T cell subpopulations by secreting IFN-γ or IL-4 concerning Th1 or Th2 responses (Ferrick et al., 1995; Ladel et al., 1996; Münz et al., 2005).

7. Suppressor or T regulatory cells (Tregs)

Regulatory T cells (Tregs) have been demonstrated to play a role in peritonitis in both animal models and people. Gershon and Kondo postulated over 30 years ago the concept of a dominant form of immunological tolerance including a specific population of "suppressor" T cells that work both to terminate normal immune responses and to avoid autoimmune disease (Gershon and Kondo, 1971). The number of Tregs in the omentum increases during tumor colonization and is associated with worse outcomes (Duraiswamy et al., 2013). Furthermore, $FoxP3^+$ Tregs expressed significant amounts of HLA-DR, CD39, and CTLA4, indicating that they are highly active in omental metastases (Kovacsovics-Bankowski et al., 2014). Tregs are also common in ovarian cancer patients' malignant ascites (Curiel et al., 2004), where their suppressive actions are aided in part by IL-6. Tregs from ascites have high levels of CTLA4 and Ki67 (Landskron et al., 2015) and have a strong immunoregulatory activity (Kampan et al., 2017).

8. Constitutive Tregs

Tregs have a $CD4^+$ cell surface and co-express CD25 activation. They constitute 5–10% of all peripheral $CD4^+$ T cells (Rathod et al., 2014; Tripathy et al., 2012b). These cells appear to develop in the thymus before spreading to the rest of the body. Positive thymic selection promotes TCRs with intermediate affinity to the MHC class II self-antigen complex on the cortical epithelium (as opposed to low affinity, which favors positive selection of conventional $CD4^+$ T cells) (Pacholczyk et al., 2002; Suto et al., 2002).

In vitro and *in vivo*, these cells have been demonstrated to limit the proliferation and cytokine production of naive $CD4^+CD25^-$ T effector cells (Rathod et al., 2014), antigen-specific Th1, Th2 cells, and $CD8^+$ T cells. The suppression process appears to be complex and includes TCR engagement, which appears to be required for effective activation and suppression. There is some evidence that Tregs impede the CTLA-4 mediated co-stimulatory signal required for T cell activation by interacting with CD80 and/or CD86 on the surface of APCs (Mills, 2004).

Decker *et al.* show that $CD4^+CD25^+$ Tregs remained relatively steady in the peripheral blood and dramatically increased in the peritoneal fluid 24 and 48 h after surgery (Decker et al., 2005), indicating that the immune response is downregulated at this stage of inflammation. Another study, which looked at the involvement of Tregs in the innate immune response, found that adoptive transfer of in vitro stimulated Tregs greatly increased survival of infected mice in a mouse model of peritonitis and sepsis. Furthermore, the benefit on survival was associated with better peritoneal bacterial clearance, increased peritoneal mast cell recruitment, and TNF-α production (Heuer et al., 2005).

Tregs that are induced: Inducible Tregs are secondary suppressor T cells that arise in the periphery from conventional $CD4^+$ T cells (CD25 negative) in response to antigens given by DCs with an activation status different from those that stimulate the development of Th1 or Th2 cells. In contrast to constitutive Tregs, their suppressive mechanism is not dependent on interaction and is mostly based on cytokine production. Tr1 and Th3 cells are the names given to the two forms of secondary Tregs. Tr1 cells are distinguished by their ability to produce high levels of IL-10 while producing low-to-moderate levels of TGF-β, whereas Th3 cells preferentially create

TGF-β. These immunosuppressive cytokines either directly or indirectly inhibit the proliferation and cytokine production of effector T cells, including Th1 cells, Th2 cells, and CTLs, by inhibiting the maturation and activation of DCs or other APCs (Jonuleit and Schmitt, 2003; Mills, 2004).

9. Cancers and peritoneum T cells

Peritoneal carcinomatosis refers to cancers that have spread to the lining surfaces of the peritoneal (abdominal) cavity from ovarian cancer, primary colorectal cancer, appendiceal cancer, or mesothelioma. Whether peritoneal cancer begins in the peritoneum or spreads from elsewhere (colon, appendix, ovary), it is classified as advanced (stage IV) once it has spread to the peritoneum and is also known as peritoneal carcinomatosis.

Peritoneal surface malignancies are a diverse set of original tumors, including peritoneal mesothelioma, as well as peritoneal metastases from other cancers, such as ovarian, gastric, colorectal, appendicular, or pancreatic cancer. Peritoneal malignancy pathogenesis is complex and not fully understood. Peritoneal carcinoma is a relatively uncommon malignancy. It begins in a thin layer of tissue lining the abdomen. The uterus, bladder, and rectum are also covered. The peritoneum is a structure made of epithelial cells. It secretes a fluid that aids in the smooth movement of organs within the abdomen. Peritoneal cancer is not the same as stomach or intestinal cancer. It is also not to be confused with tumors that spread to the peritoneum (metastasize). Peritoneal cancer develops in the peritoneum and is thus referred to as primary peritoneal cancer.

10. Cell interactions in the peritoneal cavity and omentum

Homeostasis: In the omentum, stromal cells such as peritoneal mesothelial cells and fibroblastic stromal cells assist immune cells by producing chemicals like as CXCL13 and RA, which both attract leukocytes while preserving regulatory properties.

Cancer: Cancer develops when tumor cells in the peritoneum spread to the milky areas of the omentum and change the local stromal cell populations. use adipocytes as a source of energy and increase M2 Immunity is reduced by MDSCs and Tregs, which cause polarization and accumulation of macrophages, Fig. 6 (detailed reviewed in Silva-Sanchez et al., 2021).

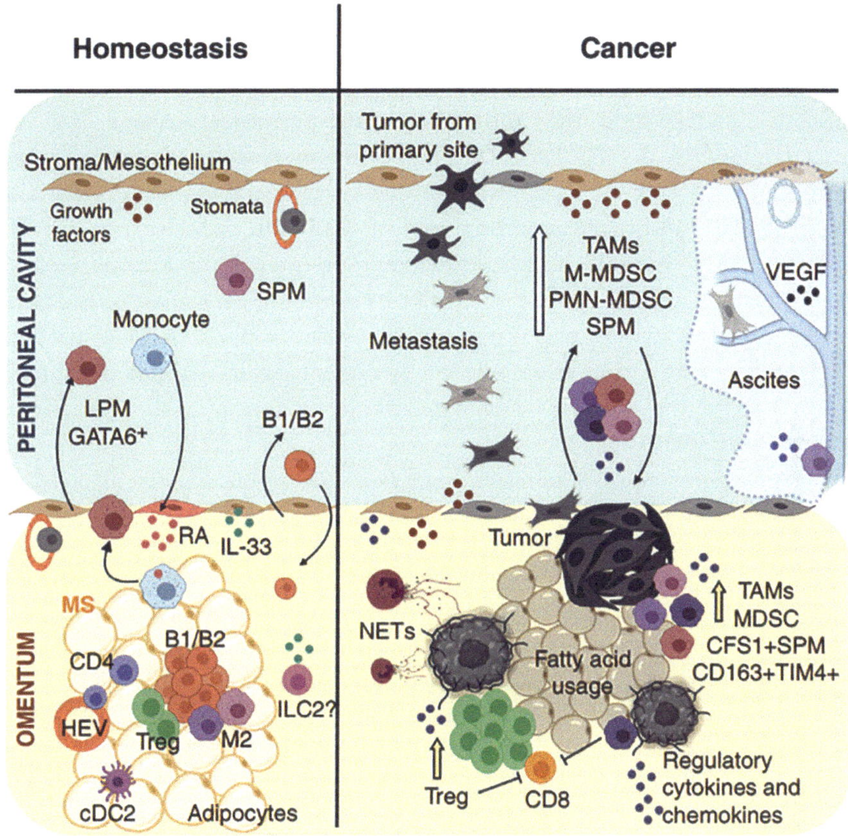

Fig. 6 Cell interactions in the peritoneal cavity and omentum.

11. Concluding remarks

This chapter highlights our current understanding of T cell and subset biology in the peritoneum. Despite significant advances in the field of adaptive immunity in recent decades, we still know very little about the unique function of T cells in the peritoneum and their role in peritoneal illnesses. The peritoneum and omentum form a unique partnership to maintain an anti-inflammatory environment in which innate immune cells, regulatory cells, and fibroblastic stroma signals serve as the first line of defense against pathogens or damage, as well as to maintain energy and metabolic processes. While this regulatory milieu avoids inflammation in steady-state conditions, it may also inhibit functional immune responses to tumor-associated Ags.

Little is known regarding the role of CTLs in peritoneal degeneration, in particular. When these cells are triggered, either by asymptomatic peritoneal viral infections or by cellular protein modification caused by a component of peritoneal dialysis fluid, they most likely cause peritoneal damage. It is also crucial to identify the regulatory T cell populations and their functions in the peritoneum. A lack of Treg response caused by renal depletion may also result in CTL autoreactivity. The role of Th3 cells, which preferentially generate TGF-, in peritoneal fibrosis would be interesting. Future research should help to understand the particular immunological milieu of the peritoneal cavity and use this knowledge to develop drugs that prevent unwanted inflammation while selectively enhancing anti-tumor immunity.

Acknowledgments

All the figures were created in partnership with Biorender and are accessible as modifiable templates on BioRender.com.

Conflicts of interest

The authors declare no conflict of interest.

Ethics approval and consent to participate

Not applicable.

Consent for publication

Not applicable.

Author contributions

Sanjay Rathod conceptualized and wrote the review.

Funding

No funding was provided for this study.

Institutional review board statement

Not applicable.

Informed consent statement

Not applicable.

Data availability statement

Not applicable.

References

Abiko, K., Mandai, M., Hamanishi, J., Yoshioka, Y., Matsumura, N., Baba, T., Yamaguchi, K., Murakami, R., Yamamoto, A., Kharma, B., Kosaka, K., Konishi, I., 2013 Mar 15. PD-L1 on tumor cells is induced in ascites and promotes peritoneal dissemination of ovarian cancer through CTL dysfunction. Clin. Cancer Res. 19 (6), 1363–1374. https://doi.org/10.1158/1078-0432.Ccr-12-2199. 23340297. eng. Epub 2013/01/24.

Abu-Hijleh, M.F., Habbal, O.A., Moqattash, S.T., 1995 Jun. The role of the diaphragm in lymphatic absorption from the peritoneal cavity. J. Anat. 186 (Pt. 3), 453–467. 7559120. eng. Epub 1995/06/01.

Albillos, A., Lario, M., Álvarez-Mon, M., 2014. Cirrhosis-associated immune dysfunction: distinctive features and clinical relevance. J. Hepatol. 61 (6), 1385–1396.

Bailey, S.R., Nelson, M.H., Himes, R.A., Li, Z., Mehrotra, S., Paulos, C.M., 2014. Th17 cells in cancer: the ultimate identity crisis. Front. Immunol. 5, 276. https://doi.org/10.3389/fimmu.2014.00276. 24987392.

Bamias, A., Tsiatas, M.L., Kafantari, E., Liakou, C., Rodolakis, A., Voulgaris, Z., Vlahos, G., Papageorgiou, T., Tsitsilonis, O., Bamia, C., Papatheodoridis, G., Politi, E., Archimandritis, A., Antsaklis, A., Dimopoulos, M.A., 2007 Jul. Significant differences of lymphocytes isolated from ascites of patients with ovarian cancer compared to blood and tumor lymphocytes. Association of CD3+CD56+ cells with platinum resistance. Gynecol. Oncol. 106 (1), 75–81. https://doi.org/10.1016/j.ygyno.2007.02.029. 17433425. eng. Epub 2007/04/17.

Becker, T.C., Wherry, E.J., Boone, D., Murali-Krishna, K., Antia, R., Ma, A., Ahmed, R., 2002 Jun 17. Interleukin 15 is required for proliferative renewal of virus-specific memory CD8 T cells. J. Exp. Med. 195 (12), 1541–1548. https://doi.org/10.1084/jem.20020369. 12070282. eng. Epub 2002/06/19.

Beelen, R., 1991. The greater omentum: physiology and immunological concepts. Neth. J. Surg. 43 (5), 145–149.

Borzio, M., Salerno, F., Piantoni, L., Cazzaniga, M., Angeli, P., Bissoli, F., Boccia, S., Colloredo-Mels, G., Corigliano, P., Fornaciari, G., 2001. Bacterial infection in patients with advanced cirrhosis: a multicentre prospective study. Dig. Liver Dis. 33 (1), 41–48.

Böttcher, K., Rombouts, K., Saffioti, F., Roccarina, D., Rosselli, M., Hall, A., Luong, T., Tsochatzis, E.A., Thorburn, D., Pinzani, M., 2018. MAIT cells are chronically activated in patients with autoimmune liver disease and promote profibrogenic hepatic stellate cell activation. Hepatology 68 (1), 172–186.

Brauner, A., Hylander, B., Wretlind, B., 1996 Mar. Tumor necrosis factor-alpha, interleukin-1 beta, and interleukin-1 receptor antagonist in dialysate and serum from patients on continuous ambulatory peritoneal dialysis. Am. J. Kidney Dis. 27 (3), 402–408. https://doi.org/10.1016/s0272-6386(96)90364-9. 8604710. eng. Epub 1996/03/01.

Caly, W.R., Strauss, E., 1993. A prospective study of bacterial infections in patients with cirrhosis. J. Hepatol. 18 (3), 353–358.

Cameron, J.S., 1995 Oct. Host defences in continuous ambulatory peritoneal dialysis and the genesis of peritonitis. Pediatr. Nephrol. 9 (5), 647–662. https://doi.org/10.1007/bf00860966. 8580033. eng. Epub 1995/10/01.

Chiesa, S., Vigo, G., Cappa, F., Prigione, I., Pistoia, V., Verrina, E., Perfumo, F., Barbano, G., 2004 Aug. Peritoneal T cell responses can be polarized toward Th1 or Th2 in children on chronic peritoneal dialysis. Artif. Organs 28 (8), 750–752. https://doi.org/10.1111/j.1525-1594.2004.00055.x. 15270959. eng. Epub 2004/07/24.

Corbett, A.J., Eckle, S.B., Birkinshaw, R.W., Liu, L., Patel, O., Mahony, J., Chen, Z., Reantragoon, R., Meehan, B., Cao, H., 2014. T-cell activation by transitory neo-antigens derived from distinct microbial pathways. Nature 509 (7500), 361–365.

Curiel, T.J., Coukos, G., Zou, L., Alvarez, X., Cheng, P., Mottram, P., Evdemon-Hogan, M., Conejo-Garcia, J.R., Zhang, L., Burow, M., Zhu, Y., Wei, S., Kryczek, I., Daniel, B., Gordon, A., Myers, L., Lackner, A., Disis, M.L., Knutson, K.L., Chen, L., Zou, W., 2004 Sep. Specific recruitment of regulatory T cells in ovarian carcinoma fosters immune privilege and predicts reduced survival. Nat. Med. 10 (9), 942–949. https://doi.org/10.1038/nm1093. 15322536. eng. Epub 2004/08/24.

Davies, S.J., Suassuna, J., Ogg, C.S., Cameron, J.S., 1989 Oct. Activation of immunocompetent cells in the peritoneum of patients treated with CAPD. Kidney Int. 36 (4), 661–668. https://doi.org/10.1038/ki.1989.243. 2811064. eng. Epub 1989/10/01.

Decker, D., Tolba, R., Springer, W., Lauschke, H., Hirner, A., von Ruecker, A., 2005 Jun 1. Abdominal surgical interventions: local and systemic consequences for the immune system—a prospective study on elective gastrointestinal surgery. J. Surg. Res. 126 (1), 12–18. https://doi.org/10.1016/j.jss.2005.01.006. 15916969. eng. Epub 2005/05/27.

Delves, P.J., Roitt, I.M., 2000 Jul 13. The immune system. Second of two parts. N. Engl. J. Med. 343 (2), 108–117. https://doi.org/10.1056/nejm200007133430207. 10891520. eng. Epub 2000/07/13.

Douvdevani, A., Rapoport, J., Konforty, A., Argov, S., Ovnat, A., Chaimovitz, C., 1994 Oct. Human peritoneal mesothelial cells synthesize IL-1 alpha and beta. Kidney Int. 46 (4), 993–1001. https://doi.org/10.1038/ki.1994.359. 7861725. eng. Epub 1994/10/01.

Duraiswamy, J., Freeman, G.J., Coukos, G., 2013 Dec 1. Therapeutic PD-1 pathway blockade augments with other modalities of immunotherapy T-cell function to prevent immune decline in ovarian cancer. Cancer Res. 73 (23), 6900–6912. https://doi.org/10.1158/0008-5472.Can-13-1550. 23975756. eng. Epub 2013/08/27.

Dusseaux, M., Martin, E., Serriari, N., Péguillet, I., Premel, V., Louis, D., Milder, M., Le Bourhis, L., Soudais, C., Treiner, E., 2011. Human MAIT cells are xenobiotic-resistant, tissue-targeted, CD161hi IL-17–secreting T cells. Blood 117 (4), 1250–1259.

Erickson, L.D., Foy, T.M., Waldschmidt, T.J., 2001. Murine B1 B cells require IL-5 for optimal T cell-dependent activation. J. Immunol. 166 (3), 1531–1539.

Ferrick, D.A., Schrenzel, M.D., Mulvania, T., Hsieh, B., Ferlin, W.G., Lepper, H., 1995 Jan 19. Differential production of interferon-gamma and interleukin-4 in response to Th1- and Th2-stimulating pathogens by gamma delta T cells in vivo. Nature 373 (6511), 255–257. https://doi.org/10.1038/373255a0. 7816142. eng. Epub 1995/01/19.

Flies, D.B., Higuchi, T., Harris, J.C., Jha, V., Gimotty, P.A., Adams, S.F., 2016 Aug. Immune checkpoint blockade reveals the stimulatory capacity of tumor-associated CD103(+) dendritic cells in late-stage ovarian cancer. Oncoimmunology 5 (8), e1185583. https://doi.org/10.1080/2162402x.2016.1185583. 27622059. eng. Epub 2016/09/14.

Ganguly, R., Milutinovich, J., Lazzell, V., Waldman, R.H., 1980 Mar. Studies of human peritoneal cells: a normal saline lavage technique for the isolation and characterization of cells from peritoneal dialysis patients. J. Reticuloendothel. Soc. 27 (3), 303–310. 7365735. eng. Epub 1980/03/01.

Gershon, R.K., Kondo, K., 1971 Dec. Infectious immunological tolerance. Immunology 21 (6), 903–914. 4943147. eng. Epub 1971/12/01.

Ghosn, E.E.B., Cassado, A.A., Govoni, G.R., Fukuhara, T., Yang, Y., Monack, D.M., Bortoluci, K.R., Almeida, S.R., Herzenberg, L.A., Herzenberg, L.A., 2010. Two physically, functionally, and developmentally distinct peritoneal macrophage subsets. Proc. Natl. Acad. Sci. U. S. A. 107 (6), 2568–2573.

Glik, A., Douvdevani, A., 2006 Jul-Aug. T lymphocytes: the "cellular" arm of acquired immunity in the peritoneum. Perit. Dial. Int. 26 (4), 438–448. 16881338. eng. Epub 2006/08/03.

Godfrey, D.I., Uldrich, A.P., McCluskey, J., Rossjohn, J., Moody, D.B., 2015. The burgeoning family of unconventional T cells. Nat. Immunol. 16 (11), 1114–1123.
Goldstein, C.S., Bomalaski, J.S., Zurier, R.B., Neilson, E.G., Douglas, S.D., 1984 Nov. Analysis of peritoneal macrophages in continuous ambulatory peritoneal dialysis patients. Kidney Int. 26 (5), 733–740. https://doi.org/10.1038/ki.1984.209. 6596459. eng. Epub 1984/11/01.
Hall, J.C., Heel, K.A., Papadimitriou, J.M., Platell, C., 1998. The pathobiology of peritonitis. Gastroenterology 114 (1), 185–196.
Hardy, R.R., 2006. B-1 B cells: development, selection, natural autoantibody and leukemia. Curr. Opin. Immunol. 18 (5), 547–555.
Hastings, W.D., Tumang, J.R., Behrens, T.W., Rothstein, T.L., 2006. Peritoneal B-2 cells comprise a distinct B-2 cell population with B-1b-like characteristics. Eur. J. Immunol. 36 (5), 1114–1123.
Haurum, J.S., Høier, I.B., Arsequell, G., Neisig, A., Valencia, G., Zeuthen, J., Neefjes, J., Elliott, T., 1999 Jul 5. Presentation of cytosolic glycosylated peptides by human class I major histocompatibility complex molecules in vivo. J. Exp. Med. 190 (1), 145–150. https://doi.org/10.1084/jem.190.1.145. 10429679. eng. Epub 1999/08/03.
Healy, J., Reznek, R., 1998. The peritoneum, mesenteries and omenta: normal anatomy and pathological processes. Eur. Radiol. 8 (6), 886–900.
Heel, K.A., Hall, J.C., 1996 Aug. Peritoneal defences and peritoneum-associated lymphoid tissue. Br. J. Surg. 83 (8), 1031–1036. https://doi.org/10.1002/bjs.1800830804. 8869299. eng. Epub 1996/08/01.
Hegde, P., Weiss, E., Paradis, V., Wan, J., Mabire, M., Sukriti, S., Rautou, P.-E., Albuquerque, M., Picq, O., Gupta, A.C., 2018. Mucosal-associated invariant T cells are a profibrogenic immune cell population in the liver. Nat. Commun. 9 (1), 1–12.
Heuer, J.G., Zhang, T., Zhao, J., Ding, C., Cramer, M., Justen, K.L., Vonderfecht, S.L., Na, S., 2005 Jun 1. Adoptive transfer of in vitro-stimulated CD4+CD25+ regulatory T cells increases bacterial clearance and improves survival in polymicrobial sepsis. J. Immunol. 174 (11), 7141–7146. https://doi.org/10.4049/jimmunol.174.11.7141. 15905557. eng. Epub 2005/05/21.
Hurst, S.M., Wilkinson, T.S., McLoughlin, R.M., Jones, S., Horiuchi, S., Yamamoto, N., Rose-John, S., Fuller, G.M., Topley, N., Jones, S.A., 2001 Jun. Il-6 and its soluble receptor orchestrate a temporal switch in the pattern of leukocyte recruitment seen during acute inflammation. Immunity 14 (6), 705–714. https://doi.org/10.1016/s1074-7613(01)00151-0. 11420041. eng. Epub 2001/06/23.
Janeway Jr., C.A., Medzhitov, R., 2002. Innate immune recognition. Annu. Rev. Immunol. 20, 197–216. https://doi.org/10.1146/annurev.immunol.20.083001.084359. 11861602. eng. Epub 2002/02/28.
Jayaprakash, P., Ai, M., Liu, A., Budhani, P., Bartkowiak, T., Sheng, J., Ager, C., Nicholas, C., Jaiswal, A.R., Sun, Y., Shah, K., Balasubramanyam, S., Li, N., Wang, G., Ning, J., Zal, A., Zal, T., Curran, M.A., 2018 Nov 1. Targeted hypoxia reduction restores T cell infiltration and sensitizes prostate cancer to immunotherapy. J. Clin. Invest. 128 (11), 5137–5149. https://doi.org/10.1172/jci96268. 30188869. eng. Epub 2018/09/07.
Jesteadt, E., Zhang, I., Yu, H., Meierovics, A., Chua Yankelevich, W.-J., Cowley, S., 2018. Interleukin-18 is critical for mucosa-associated invariant T cell gamma interferon responses to Francisella species in vitro but not in vivo. Infect. Immun. 86 (5), e00117–e00118.
Jonjić, N., Peri, G., Bernasconi, S., Sciacca, F.L., Colotta, F., Pelicci, P., Lanfrancone, L., Mantovani, A., 1992 Oct 1. Expression of adhesion molecules and chemotactic cytokines in cultured human mesothelial cells. J. Exp. Med. 176 (4), 1165–1174. https://doi.org/10.1084/jem.176.4.1165. 1383376. eng. Epub 1992/10/01.

Jonuleit, H., Schmitt, E., 2003 Dec 15. The regulatory T cell family: distinct subsets and their interrelations. J. Immunol. 171 (12), 6323–6327. https://doi.org/10.4049/jimmunol.171.12.6323. 14662827. eng. Epub 2003/12/10.

Kampan, N.C., Madondo, M.T., OM, M.N., Stephens, A.N., Quinn, M.A., Plebanski, M., 2017. Interleukin 6 present in inflammatory ascites from advanced epithelial ovarian cancer patients promotes tumor necrosis factor receptor 2-expressing regulatory T cells [Original Research]. Front. Immunol. 2017, 8. https://doi.org/10.3389/fimmu.2017.01482.

Kapsenberg, M.L., 2003 Dec. Dendritic-cell control of pathogen-driven T-cell polarization. Nat. Rev. Immunol. 3 (12), 984–993. https://doi.org/10.1038/nri1246. 14647480. eng. Epub 2003/12/04.

Kjer-Nielsen, L., Patel, O., Corbett, A.J., Le Nours, J., Meehan, B., Liu, L., Bhati, M., Chen, Z., Kostenko, L., Reantragoon, R., 2012. MR1 presents microbial vitamin B metabolites to MAIT cells. Nature 491 (7426), 717–723.

Kovacsovics-Bankowski, M., Chisholm, L., Vercellini, J., Tucker, C.G., Montler, R., Haley, D., Newell, P., Ma, J., Tseng, P., Wolf, R., Vetto, J.T., Hammill, C., Hansen, P., Weinberg, A.D., 2014. Detailed characterization of tumor infiltrating lymphocytes in two distinct human solid malignancies show phenotypic similarities. J. Immunother. Cancer 2 (1), 38. https://doi.org/10.1186/s40425-014-0038-9. 25436113. eng. Epub 2014/12/02.

Kuhns, M.S., Badgandi, H.B., 2012. Piecing together the family portrait of TCR-CD 3 complexes. Immunol. Rev. 250 (1), 120–143.

Ladel, C.H., Blum, C., Kaufmann, S.H., 1996 May. Control of natural killer cell-mediated innate resistance against the intracellular pathogen Listeria monocytogenes by gamma/delta T lymphocytes. Infect. Immun. 64 (5), 1744–1749. https://doi.org/10.1128/iai.64.5.1744-1749.1996. 8613386. eng. Epub 1996/05/01.

Lamperi, S., Carozzi, S., 1988 Mar. Interferon-gamma (IFN-gamma) as in vitro enhancing factor of peritoneal macrophage defective bactericidal activity during continuous ambulatory peritoneal dialysis (CAPD). Am. J. Kidney Dis. 11 (3), 225–230. https://doi.org/10.1016/s0272-6386(88)80154-9. 3125740. eng. Epub 1988/03/01.

Landskron, J., Helland, Ø., Torgersen, K.M., Aandahl, E.M., Gjertsen, B.T., Bjørge, L., Taskén, K., 2015 Mar. Activated regulatory and memory T-cells accumulate in malignant ascites from ovarian carcinoma patients. Cancer Immunol. Immunother. 64 (3), 337–347. https://doi.org/10.1007/s00262-014-1636-6. 25416072. eng. Epub 2014/11/25.

Lewis, S.L., Kutvirt, S.G., Cooper, C.L., Bonner, P.N., Holmes, C.J., 1993. Characteristics of peripheral and peritoneal lymphocytes from continuous ambulatory peritoneal dialysis patients. Perit. Dial. Int. (13 Suppl. 2), S273–S277. 8399587. eng. Epub 1993/01/01.

Li, F.K., Davenport, A., Robson, R.L., Loetscher, P., Rothlein, R., Williams, J.D., Topley, N., 1998 Dec. Leukocyte migration across human peritoneal mesothelial cells is dependent on directed chemokine secretion and ICAM-1 expression. Kidney Int. 54 (6), 2170–2183. https://doi.org/10.1046/j.1523-1755.1998.00174.x. 9853284. eng. Epub 1998/12/16.

Lieber, S., Reinartz, S., Raifer, H., Finkernagel, F., Dreyer, T., Bronger, H., Jansen, J.M., Wagner, U., Worzfeld, T., Müller, R., Huber, M., 2018. Prognosis of ovarian cancer is associated with effector memory CD8(+) T cell accumulation in ascites, CXCL9 levels and activation-triggered signal transduction in T cells. Oncoimmunology 7 (5), e1424672. https://doi.org/10.1080/2162402x.2018.1424672. 29721385. eng. Epub 2018/05/04.

Liu, M., Silva-Sanchez, A., Randall, T.D., Meza-Perez, S., 2021 Apr. Specialized immune responses in the peritoneal cavity and omentum. J. Leukoc. Biol. 109 (4), 717–729. https://doi.org/10.1002/jlb.5mir0720-271rr. 32881077. eng. Epub 2020/09/04.

Liuzzi, A.R., Kift-Morgan, A., Lopez-Anton, M., Friberg, I.M., Zhang, J., Brook, A.C., Roberts, G.W., Donovan, K.L., Colmont, C.S., Toleman, M.A., Bowen, T., Johnson, D.W., Topley, N., Moser, B., Fraser, D.J., Eberl, M., 2016 Sep 15. Unconventional human T cells accumulate at the site of infection in response to microbial ligands and induce local tissue remodeling. J. Immunol. 197 (6), 2195–2207. https://doi.org/10.4049/jimmunol.1600990. 27527598. eng. Epub 2016/08/17.

Lysaght, J., Allott, E.H., Donohoe, C.L., Howard, J.M., Pidgeon, G.P., Reynolds, J.V., 2011 Jul. T lymphocyte activation in visceral adipose tissue of patients with oesophageal adenocarcinoma. Br. J. Surg. 98 (7), 964–974. https://doi.org/10.1002/bjs.7498. 21520028. eng. Epub 2011/04/27.

Mariuzza, R.A., Agnihotri, P., Orban, J., 2020. The structural basis of T-cell receptor (TCR) activation: An enduring enigma. J. Biol. Chem. 295 (4), 914–925. https://doi.org/10.1074/jbc.REV119.009411. 31848223. eng. Epub 2019/12/17.

Meyers, M.A., Oliphant, M., Berne, A.S., Feldberg, M.A., 1987 Jun. The peritoneal ligaments and mesenteries: pathways of intraabdominal spread of disease. Radiology 163 (3), 593–604. https://doi.org/10.1148/radiology.163.3.3575702. 3575702. eng. Epub 1987/06/01.

Miao, Z.F., Wang, Z.N., Zhao, T.T., Xu, Y.Y., Gao, J., Miao, F., Xu, H.M., 2014 Dec. Peritoneal milky spots serve as a hypoxic niche and favor gastric cancer stem/progenitor cell peritoneal dissemination through hypoxia-inducible factor 1α. Stem Cells 32 (12), 3062–3074. https://doi.org/10.1002/stem.1816. 25142304. eng. Epub 2014/08/22.

Mills, K.H., 2004 Nov. Regulatory T cells: friend or foe in immunity to infection? Nat. Rev. Immunol. 4 (11), 841–855. https://doi.org/10.1038/nri1485. 15516964. eng. Epub 2004/11/02.

Münz, C., Steinman, R.M., Fujii, S., 2005 Jul 18. Dendritic cell maturation by innate lymphocytes: coordinated stimulation of innate and adaptive immunity. J. Exp. Med. 202 (2), 203–207. https://doi.org/10.1084/jem.20050810. 16027234. eng. Epub 2005/07/20.

Nakano, M., Ito, M., Tanaka, R., Yamaguchi, K., Ariyama, H., Mitsugi, K., Yoshihiro, T., Ohmura, H., Tsuruta, N., Hanamura, F., Sagara, K., Okumura, Y., Nio, K., Tsuchihashi, K., Arita, S., Kusaba, H., Akashi, K., Baba, E., 2018 Sep. PD-1+ TIM-3+ T cells in malignant ascites predict prognosis of gastrointestinal cancer. Cancer Sci. 109 (9), 2986–2992. https://doi.org/10.1111/cas.13723. 30187676. eng. Epub 2018/09/07.

Pacholczyk, R., Kraj, P., Ignatowicz, L., 2002. Peptide specificity of thymic selection of CD4⁺CD25⁺ T Cells. J. Immunol. 168 (2), 613–620. https://doi.org/10.4049/jimmunol.168.2.613.

Peter, J., Frey, O., Stallmach, A., Bruns, T., 2013. Attenuated antigen-specific T cell responses in cirrhosis are accompanied by elevated serum interleukin-10 levels and down-regulation of HLA-DR on monocytes. BMC Gastroenterol. 13 (1), 1–10.

Platell, C., Cooper, D., Papadimitriou, J.M., Hall, J.C., 2000. The omentum. World J. Gastroenterol. 6 (2), 169–176. https://doi.org/10.3748/wjg.v6.i2.169. 11819552.

Rangel-Moreno, J., Moyron-Quiroz, J.E., Carragher, D.M., Kusser, K., Hartson, L., Moquin, A., Randall, T.D., 2009. Omental milky spots develop in the absence of lymphoid tissue-inducer cells and support B and T cell responses to peritoneal antigens. Immunity 30 (5), 731–743.

Rathod, S., 2022. Novel insights into the immunotherapy-based treatment strategy for autoimmune Type 1 diabetes. Diabetology 3 (1), 79–96. https://doi.org/10.3390/diabetology3010007.

Rathod, S.B., Das, R., Thanapati, S., Arankalle, V.A., Tripathy, A.S., 2014 Feb. Suppressive activity and altered conventional phenotype markers/mediators of regulatory T cells in patients with self-limiting hepatitis E. J. Viral Hepat. 21 (2), 141–151. https://doi.org/10.1111/jvh.12125. 24383927. eng. Epub 2014/01/05.

Riera-Domingo, C., Audigé, A., Granja, S., Cheng, W.C., Ho, P.C., Baltazar, F., Stockmann, C., Mazzone, M., 2020 Jan 1. Immunity, hypoxia, and metabolism-the ménage à trois of cancer: implications for immunotherapy. Physiol. Rev. 100 (1), 1–102. https://doi.org/10.1152/physrev.00018.2019. 31414610. eng. Epub 2019/08/16.

Rissoan, M.C., Soumelis, V., Kadowaki, N., Grouard, G., Briere, F., de Waal, M.R., Liu, Y.J., 1999 Feb 19. Reciprocal control of T helper cell and dendritic cell differentiation. Science 283 (5405), 1183–1186. https://doi.org/10.1126/science.283.5405.1183. 10024247. eng. Epub 1999/02/19.

Riva, A., Patel, V., Kurioka, A., Jeffery, H.C., Wright, G., Tarff, S., Shawcross, D., Ryan, J.M., Evans, A., Azarian, S., 2018. Mucosa-associated invariant T cells link intestinal immunity with antibacterial immune defects in alcoholic liver disease. Gut 67 (5), 918–930.

Russ, M.A., Reith, H.B., 2003 Fall. The severity of infection induces a shift in the type 1/Type 2 T-helper cell balance in patients with or without peritonitis. Surg. Infect. (Larchmt.) 4 (3), 247–254. https://doi.org/10.1089/109629603322419580. 14588159. eng. Epub 2003/11/01.

Saccani, A., Schioppa, T., Porta, C., Biswas, S.K., Nebuloni, M., Vago, L., Bottazzi, B., Colombo, M.P., Mantovani, A., Sica, A., 2006 Dec 1. p50 nuclear factor-kappaB overexpression in tumor-associated macrophages inhibits M1 inflammatory responses and antitumor resistance. Cancer Res. 66 (23), 11432–11440. https://doi.org/10.1158/0008-5472.Can-06-1867. 17145890. eng. Epub 2003/11/01.

Scharping, N.E., Menk, A.V., Whetstone, R.D., Zeng, X., Delgoffe, G.M., 2017 Jan. Efficacy of PD-1 blockade is potentiated by metformin-induced reduction of tumor hypoxia. Cancer Immunol. Res. 5 (1), 9–16. https://doi.org/10.1158/2326-6066.Cir-16-0103. 27941003. eng. Epub 2016/12/13.

Schwenger, V., Morath, C., Salava, A., Amann, K., Seregin, Y., Deppisch, R., Ritz, E., Bierhaus, A., Nawroth, P.P., Zeier, M., 2006 Jan. Damage to the peritoneal membrane by glucose degradation products is mediated by the receptor for advanced glycation end-products. J. Am. Soc. Nephrol. 17 (1), 199–207. https://doi.org/10.1681/asn.2005020155. 16319192. eng. Epub 2005/12/02.

Seder, R.A., Gazzinelli, R., Sher, A., Paul, W.E., 1993 Nov 1. Interleukin 12 acts directly on CD4+ T cells to enhance priming for interferon gamma production and diminishes interleukin 4 inhibition of such priming. Proc. Natl. Acad. Sci. U. S. A. 90 (21), 10188–10192. https://doi.org/10.1073/pnas.90.21.10188. 7901851. eng. Epub 1993/11/01.

Selin, L.K., Santolucito, P.A., Pinto, A.K., Szomolanyi-Tsuda, E., Welsh, R.M., 2001 Jun 1. Innate immunity to viruses: control of vaccinia virus infection by gamma delta T cells. J. Immunol. 166 (11), 6784–6794. https://doi.org/10.4049/jimmunol.166.11.6784. 11359837. eng. Epub 2001/05/22.

Shedlock, D.J., Shen, H., 2003 Apr 11. Requirement for CD4 T cell help in generating functional CD8 T cell memory. Science 300 (5617), 337–339. https://doi.org/10.1126/science.1082305. 12690201. eng. Epub 2003/04/12.

Shipley, P., Cunningham, R., 1916. STUDIES ON ABSORPTION FROM SEROUS CAVITIES: I. The Omentum as a Factor in Absorption from the Peritoneal Cavity. Am. J. Physiol. 40 (1), 75–81.

Suto, A., Nakajima, H., Ikeda, K., Kubo, S., Nakayama, T., Taniguchi, M., Saito, Y., Iwamoto, I., 2002 Jan 15. CD4(+)CD25(+) T-cell development is regulated by at least 2 distinct mechanisms. Blood 99 (2), 555–560. https://doi.org/10.1182/blood.v99.2.555. 11781238. eng. Epub 2002/01/10.

Swain, S.L., Weinberg, A.D., English, M., Huston, G., 1990 Dec 1. IL-4 directs the development of Th2-like helper effectors. J. Immunol. 145 (11), 3796–3806. 2147202. eng. Epub 1990/12/01.

Swain, S.L., McKinstry, K.K., Strutt, T.M., 2012. Expanding roles for CD4+ T cells in immunity to viruses. Nat. Rev. Immunol. 12 (2), 136–148.

Szender, J.B., Emmons, T., Morrell, K., Belliotti, S., Dickson, D., Khan, A., Mayor, P.C., Lele, S.B., Frederick, P.J., Zsiros, E., 2017. Impact of ascites volume on clinical outcomes in epithelial ovarian cancer: A cohort study. Gynecol. Oncol. 145, 79.

Tauer, A., Zhang, X., Schaub, T.P., Zimmeck, T., Niwa, T., Passlick-Deetjen, J., Pischetsrieder, M., 2003 Mar. Formation of advanced glycation end products during CAPD. Am. J. Kidney Dis. 41 (3 Suppl. 1), S57–S60. https://doi.org/10.1053/ajkd.2003.50086. 12612954. eng. Epub 2003/03/04.

Terri, M., Trionfetti, F., Montaldo, C., Cordani, M., Tripodi, M., Lopez-Cabrera, M., Strippoli, R., 2021. Mechanisms of peritoneal fibrosis: focus on immune cells-peritoneal stroma interactions. Front. Immunol. 12, 607204. https://doi.org/10.3389/fimmu.2021.607204. 33854496.

Topley, N., Jörres, A., Luttmann, W., Petersen, M.M., Lang, M.J., Thierauch, K.H., Müller, C., Coles, G.A., Davies, M., Williams, J.D., 1993 Jan. Human peritoneal mesothelial cells synthesize interleukin-6: induction by IL-1 beta and TNF alpha. Kidney Int. 43 (1), 226–233. https://doi.org/10.1038/ki.1993.36. 8433563. eng. Epub 1993/01/01.

Tripathy, A.S., Das, R., Rathod, S.B., Arankalle, V.A., 2012a. Cytokine profiles, CTL response and T cell frequencies in the peripheral blood of acute patients and individuals recovered from hepatitis E infection. PLoS One 7 (2), e31822. https://doi.org/10.1371/journal.pone.0031822. 22384080. eng. Epub 2012/03/03.

Tripathy, A.S., Das, R., Rathod, S.B., Gurav, Y.K., Arankalle, V.A., 2012b. Peripheral T regulatory cells and cytokines in hepatitis E infection. Eur. J. Clin. Microbiol. Infect. Dis. 31 (2), 179–184. https://doi.org/10.1007/s10096-011-1291-1.

Valle, M.T., Degl'Innocenti, M.L., Giordano, P., Kunkl, A., Costantini, M.T., Perfumo, F., Manca, F., Gusmano, R., 1989 Nov. Analysis of cellular populations in peritoneal effluents of children on CAPD. Clin. Nephrol. 32 (5), 235–238. 2582649. eng. Epub 1989/11/01.

Wang, H.H., Lin, C.Y., 2005 Aug. Interleukin-12 and -18 levels in peritoneal dialysate effluent correlate with the outcome of peritonitis in patients undergoing peritoneal dialysis: implications for the Type I/Type II T-cell immune response. Am. J. Kidney Dis. 46 (2), 328–338. https://doi.org/10.1053/j.ajkd.2005.05.008. 16112053. eng. Epub 2005/08/23.

Wang, R., Dillon, C.P., Shi, L.Z., Milasta, S., Carter, R., Finkelstein, D., McCormick, L.L., Fitzgerald, P., Chi, H., Munger, J., 2011. The transcription factor Myc controls metabolic reprogramming upon T lymphocyte activation. Immunity 35 (6), 871–882.

Wefers, C., Duiveman-de Boer, T., Zusterzeel, P.L.M., Massuger, L., Fuchs, D., Torensma, R., Wheelock, C.E., de Vries, I.J.M., 2018 Jan 17. Different lipid regulation in ovarian cancer: inhibition of the immune system. Int. J. Mol. Sci. 19 (1). https://doi.org/10.3390/ijms19010273. 29342108. eng. Epub 2018/01/18.

Wiest, R., Lawson, M., Geuking, M., 2014. Pathological bacterial translocation in liver cirrhosis. J. Hepatol. 60 (1), 197–209.

Wucherpfennig, K.W., Gagnon, E., Call, M.J., Huseby, E.S., Call, M.E., 2010. Structural biology of the T-cell receptor: insights into receptor assembly, ligand recognition, and initiation of signaling. Cold Spring Harb. Perspect. Biol. 2 (4), a005140.

Zeillemaker, A.M., Mul, F.P., Hoynck van Papendrecht, A.A., Kuijpers, T.W., Roos, D., Leguit, P., Verbrugh, H.A., 1995 Feb. Polarized secretion of interleukin-8 by human mesothelial cells: a role in neutrophil migration. Immunology 84 (2), 227–232. 7750998. eng. Epub 1995/02/01.

Ziegler, H.K., 2004. The role of gamma/delta T cells in immunity to infection and regulation of inflammation. Immunol. Res. 29 (1–3), 293–302. https://doi.org/10.1385/ir:29:1-3:293. 15181290. eng. Epub 2004/06/08.

CHAPTER THREE

Innate immune response to peritoneal bacterial infection

Cristina Català[a], María Velasco-de Andrés[a], Sergi Casadó-Llombart[a], Alejandra Leyton-Pereira[a], Laura Carrillo-Serradell[a], Marcos Isamat[b], and Francisco Lozano[a,c,d,*]

[a]Institut d'Investigacions Biomèdiques August Pi i Sunyer (IDIBAPS), Barcelona, Spain
[b]Sepsia Therapeutics S.L. 08908 L'Hospitalet de Llobregat, Barcelona, Spain
[c]Servei d'Immunologia, Centre de Diagnòstic Biomèdic (CDB), Hospital Clínic de Barcelona, Barcelona, Spain
[d]Departament de Biomedicina, Facultat de Medicina, Universitat de Barcelona, Barcelona, Spain
*Corresponding author: e-mail address: flozano@clinic.cat

Contents

Abstract

Spontaneous and secondary peritoneal infections, mostly of bacterial origin, easily spread to cause severe sepsis. Cellular and humoral elements of the innate immune system are constitutively present in peritoneal cavity and omentum, and play an important role in peritonitis progression and resolution. This review will focus on the description of the anatomic characteristics of the peritoneal cavity and the composition and function of such innate immune elements under both steady-state and bacterial infection conditions. Potential innate immune-based therapeutic interventions in bacterial peritonitis alternative or adjunctive to classical antibiotic therapy will be briefly discussed.

International Review of Cell and Molecular Biology, Volume 371
ISSN 1937-6448
https://doi.org/10.1016/bs.ircmb.2022.04.014

1. Peritoneal bacterial infections

Peritonitis is an inflammation of the serous membrane which lines the intrabdominal viscera and the abdominal cavity wall, known as peritoneum. Peritonitis can result from localized or diffuse intra-abdominal infections (e.g., appendicitis or diverticulitis) but also from non-infectious processes (e.g., pancreatitis, autoimmune serositis or peritoneal carcinomatosis). In the first case, most intra-abdominal infections develop in the gastrointestinal tract from an aerobic or anaerobic enteric bacterial source. These infections are characterized by a rapid migration of the microorganisms from the peritoneal cavity into systemic circulation. Therefore, they are associated with important morbidity and mortality due to secondary multiorgan dysfunction. Peritoneal infections are usually classified as primary, secondary and tertiary based on their origin and disease progression.

1.1 Primary peritoneal infection: Spontaneous bacterial peritonitis

A primary peritoneum infection or spontaneous bacterial peritonitis (SBP) is defined as the peritoneum infection without an easily identifiable intra-abdominal source or visceral defect. It typically occurs in cirrhotic patients but it can arise as a result of ascites fluid accumulation from other pathologies, such as liver disease, systemic lupus erythematosus or cancer (Song, 2018). Approximately 10–25% of patients with ascites will develop SBP, and the condition is associated with a 20% in-hospital rate of mortality.

The most common etiological agents are gram-negative bacteria, with a special focus on *Entercoccus* species in cirrhotic patients, *Staphylococcus aureus* in patients under peritoneal dialysis and the *Pneumoccoccus* genera in young women (Oey et al., 2018). The infection is a result of bacterial overgrowth in combination with intestinal barrier damage, local immune system alterations as well as slow bowel motility in cirrhotic patients and lower opsonization ability (Koulaouzidis et al., 2007).

1.2 Secondary peritoneal infection

Secondary peritonitis is defined as the peritoneum inflammation resulting from a direct contact with a peritoneal contaminant. The typical pathological condition develops from the disruption of gastrointestinal (GI) tract integrity and subsequently presents a polymicrobial etiology. Aerobic and anaerobic facultative gram-negative enteric bacteria, including *Escherichia coli*, *Klebsiella* sp. and *Pseudomonas* sp. are typically isolated from GI perforations. Etiologies usually include acute perforation, necrosis, cancer, or organ inflammation among others (Clements et al., 2021).

1.3 Tertiary peritoneal infection

A tertiary peritoneal infection is a peritonitis which persists or relapses 48 h after clinical management of secondary peritonitis. It is typically due to altered intestinal microbiome composition, immune system deterioration or progressive organ failure (Marques et al., 2021). A microbial shift was described in tertiary peritonitis in which previous microbial species led to milder opportunistic bacteria and fungi such as *Enterococcus, Enterobacter, Pseudomonas* species or *Candida albicans* (Bass et al., 2021).

2. Anatomy of the peritoneal cavity

The peritoneum is a serous membrane that covers the abdominal cavity in amniotes. It consists of a single layer of mesothelial cells over a basement membrane, which forms a lining on most abdominal organs, including liver, spleen, stomach, bowels and the female reproductive organs (van Baal et al., 2017). It is divided into the parietal peritoneum, which covers the abdominal wall and diaphragm, and the visceral peritoneum, which covers the visceral organs. The cavity between the parietal and the visceral peritoneum constitutes the peritoneal cavity, which is the biggest cavity in human body (Fig. 1). While such cavity is located around abdominal organs, it does not contain any organ inside; it normally holds only a small amount of lubricant fluid—approximately 100 mL—secreted by the peritoneal membrane (Pannu and Oliphant, 2015). Such fluid contains surfactants as well as cells, enzymes and factors that regulate different homeostatic cell functions in the peritoneal cavity. Several pathologies may lead to an increase of abdominal capillary pressure, leading to accumulation of fluid in the peritoneal cavity, a condition known as ascites. Ascites can be the result of different pathologies, including cirrhosis, primary or metastatic peritoneal carcinomatosis, infections, and pancreatitis (Aithal et al., 2021; Carrier et al., 2014; Sangisetty and Miner, 2012).

The peritoneum is perforated by *stomata*—small fenestrations between the mesothelial cells that drain fluids, particles and cells from the peritoneal cavity into regional lymph nodes. In normal conditions, immune cells are

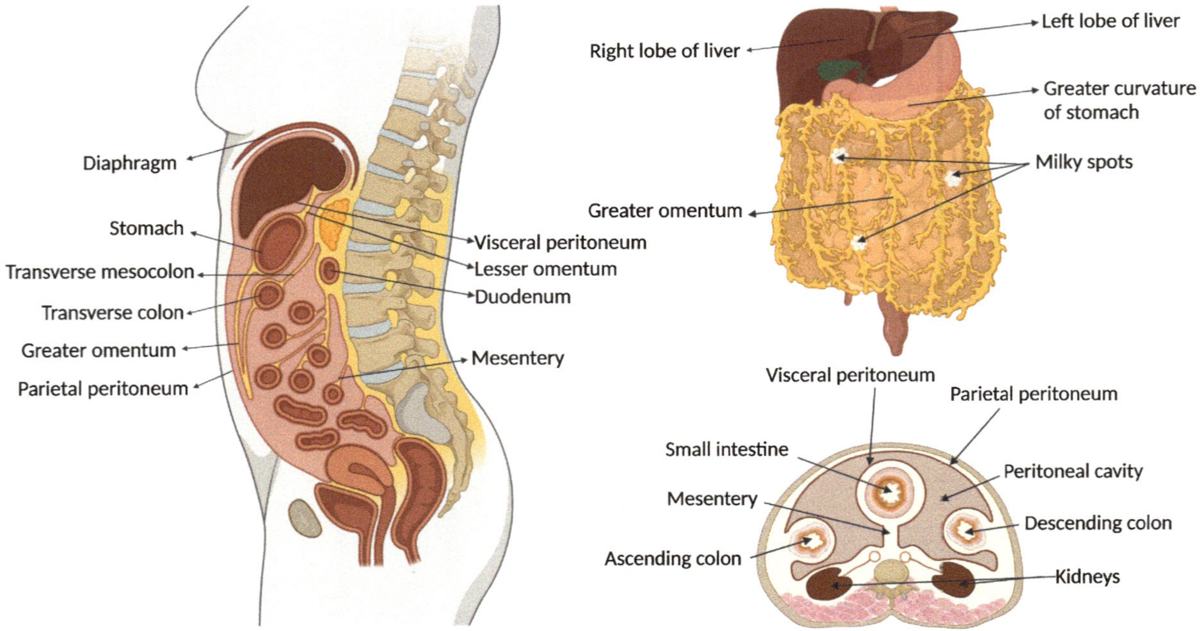

Fig. 1 Schematic anatomical view of peritoneal cavity and omentum (lateral, front and cross sections).

found in the peritoneal cavity, mainly macrophages, B1 lymphocytes and other cell types (Baumgarth, 2011; Hayakawa et al., 1985; Lewis and Holmes, 1991).

The abdominal cavity also contains portions of double layer of peritoneum that attach the organs to the abdominal wall, including the omentum and the mesentery (Fig. 1). The *lesser omentum* is bound to the liver, the lesser curvature of the stomach, and the duodenum. The *greater omentum* hangs from the stomach, the frontal side of the intestines, and extends to the diaphragm, the transverse colon and the spleen. It is highly vascularized and contains adipose tissue. The fold created by the lesser and greater omenta generates a space known as the *lesser sac* or *omental bursa*, which is topologically part of the peritoneal cavity. The remaining space of the peritoneal cavity is the *greater sac or general cavity*, and connects with the lesser sac via the omental foramen.

The *mesentery* is a contiguous structure associated with the small intestine and colon (Sehgal and Calvin Coffey, 2014), and contains a network of lymphatic vessels and nodes (Culligan et al., 2014). Mesenteric lymph nodes drain lymph from the small intestinal and colonic lymphatic network, which plays a central role in gut homeostasis, clearing of intestinal fluid and immune cells, and lipid transport (Alexander et al., 2010).

The omentum together with other adipose tissues (e.g., mesenteric, mediastinal and pericardial fat) contain lymphoid structures known as *milky spots* or fat-associated lymphoid clusters (FALCs), which have an immune composition similar to that of the secondary lymphoid tissue: a central B cell area enclosed by T and myeloid cells, supported by a fibroblast cell network and embedded between adipocytes just beneath the mesothelial cell layer (Liu et al., 2021). During inflammation, the number and size of omental milky spots dramatically increases and some develop into germinal centers and produce antibodies (Bénézech et al., 2015; Mebius, 2009; Meza-Perez and Randall, 2017; Ranvier, 1874).

2.1 Innate immune system of the peritoneal cavity and omentum

The peritoneal cavity and omentum has a complex immune system, whose function is essential for the understanding of peritoneal infection. As mentioned before, peritoneal resident leukocytes, omentum and mesentery conform the immune system in the peritoneal cavity and are the source of immune cells that will respond to pathogens and control infection. Omentum plays an early role in infection resolution, as it is evidenced in animal sepsis studies where omentectomy has detrimental effects in sepsis survival (Wang et al., 2019). However, omentum deficiency can be compensated after several days after omentectomy. Specifically, 28 days after omentectomy the peritoneal resident macrophage population is restored (Wang et al., 2019). The innate immune cell populations—both from myeloid or lymphoid origin—and soluble factors are critical in preventing peritoneal infection progression will be discussed in next sections and illustrated in Fig. 2.

2.1.1 Cell components

2.1.1.1 Stromal cells

Basic homeostatic issues such as migration, homing, survival and proliferation of immune cells of hematopoietic origin are sustained by factors provided by local stromal cells, which include mesothelial cells, fibroblasts, blood and lymphatic endothelial cells and adipocytes. This is exemplified by mesothelial and/or fibroblastic cell promotion of peritoneal macrophage (Mϕ) differentiation and IgA isotype-switching of omentum B1 cells via retinoic acid production (Okabe and Medzhitov, 2014; Roy et al., 2013), Th2/ILC2/Treg/M2 phenotype differentiation via alarmin (IL-33) production (Mahlakõiv et al., 2019), and attraction or $CCR7^+$ naïve T/B cells via CCL19 production (Perez-Shibayama et al., 2018). Upon peritoneal infection mesothelial and/or fibroblastic cells trigger cytokine and chemokine production that promote neutrophil (CXCL2/SDF-1 and CXCL5) and monocyte (CCL2) accumulation into the peritoneal cavity and omentum.

2.1.1.2 Myeloid cells

Resident peritoneal myeloid immune cells include Mϕ ($F4/80^{high}$ $GR1^-$), monocytes ($F4/80^{med}$ $GR1^+$), dendritic cells and mast cells. These cells recognize conserved microbial products—microbial associated molecular patterns (MAMPs)—through different families of pathogen recognition receptors (PRRs), including toll-like (TLRs), C-type lectin and scavenger receptors among others. Pathogen binding to these receptors leads to release of pro-inflammatory mediators (mostly cytokines and chemokines) that recruit or activate other innate immune cells, such as neutrophils and Mϕ, to the site of infection (Sutherland et al., 2008).

Mϕ derive from monocytes and have different functions, but they mostly act during homeostatic patrolling of peritoneal cavity to eliminate pathogens or dead cells. Mϕ have different surface PRR that enable them to recognize bacteria and phagocyte them, in order to kill or digest them. In addition, Mϕ

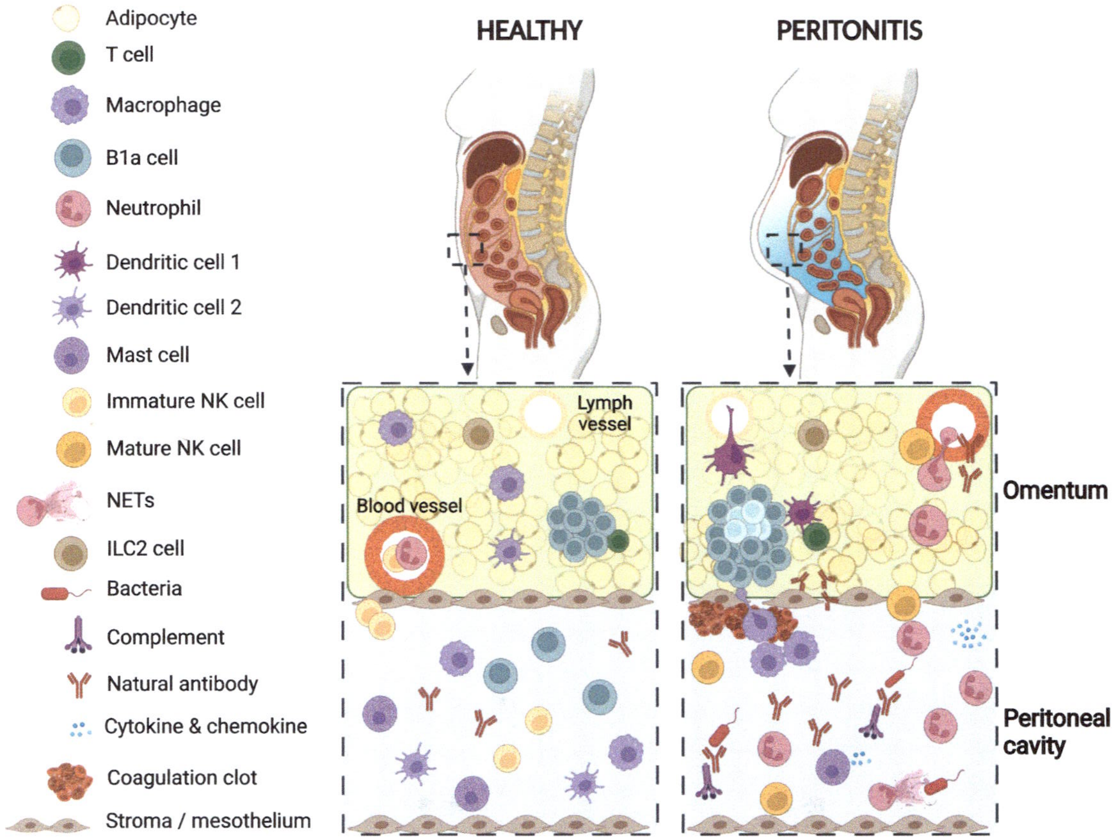

Fig. 2 Schematic representation of the cellular and humoral components of the innate immune system from the peritoneal cavity and omentum in health (left) and infection (right).

produce reactive oxygen species, such as nitric oxide, which help killing phagocytized bacteria (Gordon, 2016). They can also process and present antigens, produce chemokines and cytokines such as interleukin-1β (IL-1β), IL-6, IL-12, IL-18, IL-23, tumor necrosis factor (TNF-α), and IL-10. These mediators are responsible for neutrophil recruitment to the peritoneal cavity. During peritonitis, a special phenomenon occurs known as "macrophage disappearance reaction" (MDR). It consists on a dramatic reduction of Mϕ cell numbers that can be obtained from peritoneal lavage samples. This phenomenon is due to alterations in the coagulation cascade and cell adhesion ability that prompt Mϕ adhesion to the mesothelial wall of the peritoneal cavity, where they continue carrying out their functions (Okabe and Medzhitov, 2014; Zwicky et al., 2021).

Dendritic cells (DCs) are the main linker between innate and adaptive immunity as they are considered the best antigen presenting cells. Their major task is the production of chemokines that will recruit Mϕ, neutrophils, natural killer (NK) cells and immature DCs at the inflammatory site, and then migrate to nearby lymphoid organs to encounter antigen-specific T cells (McCully and Madrenas, 2006; De Smedt et al., 1996). However, the predominant DC population in omentum is not the conventional DC type 1 (cDC1) but cDC2 cells instead, with anti-inflammatory and repair functions, which can be subverted in the context of infection.

Mast cells are also peritoneal resident cells (Jamur et al., 2010). Prior work has established that mast cells play an essential role in regulating the immune response in acute bacterial peritonitis by secreting TNF-α and tryptase, which induce the extravasation of bactericidal granulocytes into the site to control the infection (Hültner et al., 1996; Thakurdas et al., 2007). The mast cell-specific IL-6 deficient mice in response to caecal ligation and puncture (CLP) sepsis and *K. pneumoniae* lung and peritonitis infection suggest that mast cell IL-6 is a major mediator of survival in these severe infections and that it protects from death by enhancing intracellular killing of bacteria by neutrophils (Sutherland et al., 2008).

Neutrophils provide a first line of defense of the innate immune system by phagocytosis, killing, and digesting bacteria. Neutrophils are the most abundant immune cell type in blood, where they circulate as dormant cells, and contain distinctive cytoplasmic granules. Neutrophils do not reside in the peritoneal cavity. Upon peritoneal infection, endothelial cells trap bypassing neutrophils and guide them through the endothelial cell lining whereby the neutrophils are activated and tuned for their subsequent interaction with microbes (Sawant et al., 2021). Once in the peritoneal cavity, neutrophils recognize microorganisms through different PRRs or opsonic receptors and kill them by means of microbicidal agents (e.g., reactive oxygen species, proteolytic enzymes or antibacterial peptides) released from granules or de novo generated by metabolic activation. At the same time, neutrophils can secrete pro-inflammatory cytokines (e.g., TNF-α) that contribute to acute inflammation. As a final act, neutrophils can extrude strands of DNA with bactericidal proteins attached (e.g., histones) that act as extracellular traps (NETs) for microorganisms (Borregaard, 2010; Nauseef and Borregaard, 2014; Segal, 2005). Hence, correct neutrophil infiltration into peritoneal cavity is crucial for the eradication of bacteria, prevention of infection complications and the survival of peritonitis patients (Frydrych et al., 2019).

2.1.1.3 Innate-type lymphocytes

It is now widely accepted that the immune system is a complex organization which includes different cell subsets with intermediate properties, in addition to the classical innate and adaptive systems. This is the case of the so-called innate-type lymphocytes comprising a variety of lymphoid subsets. These cells exert their function through the expression of a wide repertoire of activating and inhibitory receptors. Innate-type lymphocytes include B1a cells, natural killer (NK) cells, innate lymphoid cells (ILCs) of three subtypes (ILC1s, ILC2s and ILC3s), NKT and γδT cells, which develop different responses depending on the pathogen.

B1a cells are peritoneal cavity resident cells and are also present in the omentum. These are innate-like B lymphocytes that differ from conventional B2 cells by their ability to spontaneously secrete germline-like, repertoire-skewed, polyreactive natural antibodies (NAbs). These NAbs act as a first line of defense by neutralizing a wide range of pathogens before triggering the adaptive immune response (Aziz et al., 2015). B1a cells spontaneously secrete and maintain steady-state levels of NAbs of mainly immunoglobulin M (IgM) isotype in naïve, non-immunized, and antigen-free mice (Grönwall et al., 2012). Specifically, NAbs levels are maintained in a steady-state within the peritoneal cavity by resident B1 cell subsets. B1a cell-derived NAbs counteract a wide range of bacterial and viral infections (Baumgarth et al., 2005). NAbs interaction with an infectious agent can either exert direct inhibition, complement activation, or opsonization leading to phagocytosis and/or antibody-dependent cell-mediated cytotoxicity (ADCC) (Grönwall et al., 2012; Rothstein et al., 2013). Moreover, after endotoxin stimulation, B1a cells migrate from body cavities to the spleen where they secrete higher amounts of NAbs (Ha et al., 2006; Kawahara et al., 2003; Yang et al., 2007). In line with these data, studies performed in B cell deficient ($Rag1^{-/-}$ and $\mu MT^{-/-}$) mice support an essential role for B cells in early immune response to sepsis. Such mice show higher susceptibility to sepsis, which can be reverted by adoptive B cell transfer (Hotchkiss et al., 1999; Kelly-Scumpia et al., 2011).

Likewise, numerous studies report an important role of NK cells during bacterial infection. Mice lacking B and T cells but not NK cells show higher survival rates and lower bacterial titers than those lacking also NK cells upon *Shigella flexneri* infection (Le-Barillec et al., 2005). NK cells can recognize microorganisms through PRRs and exert anti-bacterial activity. Different bacteria species and bacterial derived-structures activate NK cells through different TLRs (Noh et al., 2020), nucleotide-binding oligomerization domain (NOD)-like receptors (NLRs) (Qiu et al., 2011), and retinoic acid inducible gene I (RIG-I)-like receptors (RLRs) (Perrot et al., 2010). NK cells also express activating and inhibitory receptors that are involved in the immune response against bacteria. This is the case of the NKp44 receptor that binds to *Pseudomonas aeruginosa* and *Mycobacteria* (Esin et al., 2008). In addition, NK cell activation requires a continuous crosstalk with DCs or Mϕ producing pro-inflammatory cytokines such us IL-12 and IL-18 (Nomura et al., 2002). Once activated, these cells can exert both direct and indirect activities. The direct anti-bacterial ability comes from the secretion of soluble molecules such as granzymes, granulysins and perforins stored in cytotoxic granules (Krensky and Clayberger, 2009; Lauw et al., 2000; Lu et al., 2014). The indirect anti-bacterial activity of NK cells relies on IFN-γ and TNF-α cytokine secretion by interaction with other immune system cells (Marcenaro et al., 2008). IFN-γ plays a crucial role against bacterial infections by inducing phagocytosis of bacteria and infected cells by Mϕ (McSharry and Gardiner, 2010). In turn, NK cells also play an important role in bacterial clearance by promoting the function of other innate immune cells such as neutrophils and Mϕ (Byrne et al., 2004; Small et al., 2008; Tsuchiya et al., 2012).

A resident NK-population sharing phenotypical characteristics with immature NK cells from spleen and liver has been described in the peritoneal cavity (Gonzaga et al., 2011), which are proficient pro-inflammatory cytokine producers with little cytotoxic activity. Furthermore, NK cells are recruited to the peritoneum in a CCR2-dependent manner after TLR2 activation (Haidl et al., 2020). Studies in NK deficient mice undergoing CLP highlighted the role of these cells in septic peritonitis (Godshall et al., 2003). Thus, NK cells exert an early anti-bacterial response, both local and systemic, and induces Mϕ activation.

Other innate-type immune cells are the innate lymphoid cells (ILCs). These are tissue-resident lymphocytes found in both lymphoid and non-lymphoid tissues (rarely in the peripheral blood) that lack adaptive antigen-specific T and B cell receptors (TCR and BCR) (Vivier et al., 2018). They are particularly abundant on mucosal surfaces, playing a key role in mucosal immunity and homeostasis. ILCs are divided into three groups: ILC1 resemble Th1 cells implicated in immunity against intracellular pathogens; ILC2 resemble Th2 cells implicated in immunity against helminths, and ILC3 resemble Th17 and Th22 cells. ILC1 and ILC3 are abundant at mucosal sites and involved in the innate immune response to extracellular bacteria and the containment of intestinal commensals (Abt et al., 2015; Buonocore et al., 2010; Rankin et al., 2016; Sonnenberg et al., 2011). As in most adipose tissues, ILC2 cells are also important resident cells in the omentum, where they facilitate B1 cell differentiation into antibody-secreting cells (Neill et al., 2010) and "beiging" of adipocytes from precursor cells (Brestoff et al., 2014).

NKT and Tγδ cells are similar to these innate lymphoid cells. In the case of NKT cells, they are present in the omentum (Lynch et al., 2009) and their ability to recognize and clear bacteria has been described (Kinjo et al., 2005) and similar functions have been attributed to Tγδ cells (Hsu et al., 2015).

2.1.2 Humoral components

The humoral components of the innate immune system also play a relevant role in the peritoneal antibacterial response. The main ones are the serine protease cascades of the complement and coagulation systems, as well as NAbs (see Section 2.1.1.2), acute phase proteins (e.g., pentraxins), cytokines and chemokines, and antimicrobial peptides.

Complement activation occurs after bacterial infection by one of three initiation pathways: classical, alternative or lectin pathways. The first one is activated when antibodies of IgM or IgG class bind to microbial antigens, while the other two can be directly activated by the microbial surfaces. Each pathway contains a C3 convertase that cleaves C3 producing C3b and subsequently a C5 convertase. C3b fragments opsonize microbial surfaces and facilitate their phagocytosis by neutrophils and Mϕ, as well as B cell activation. The C5 convertase results in C5b deposition and initiation of the common terminal pathway: the formation of the membrane attack complex (MAC), which perforates and lyses the pathogenic cells. The C3 and C5 convertases also produce potent anaphylatoxins (C3a and C5a), which promote endothelial activation, as well as neutrophil, mast cell, and Mϕ activation and mobilization to the infection site (Girardi et al., 2020). In fact, C5a anaphylotoxin is the most active inflammatory peptide produced during sepsis (Guo and Ward, 2005). Consequently, administration of anti-C5a blocking antibodies to septic rats or mice undergoing polymicrobial peritonitis following CLP is highly effective in diminishing the severity of sepsis and improving the outcome (Czermak et al., 1999; Rittirsch et al., 2008).

The *coagulation* cascade is activated when negatively charged surfaces of damaged cells induce the proteolytic cleavage of Factor XII (FXII; Hageman factor) to FXIIa. This leads to clot formation and cleavage of prekallikrein to kallikrein for subsequent release of bradykinin, which in turn induces vasodilation, neutrophil chemotaxis and vascular permeability. There is a crosstalk between the complement and coagulation cascades, so that both FXIIa and kallikrein activate the complement cascade independent of other complement initiators (Shishido et al., 2012).

Acute phase proteins are rapidly synthesized in the liver upon induction by pro-inflammatory cytokines (IL-1β, IL-6 and TNF-α) released by innate immune cells subjected to microbial aggression. This is the case of *pentraxins* that recognize and bind multiple pathogens as well as endogenous ligands, including apoptotic cells and extracellular matrix components. They include C-reactive protein (CRP) and serum amyloid P protein (SAP). CRP facilitates clearance of pathogens and cell debris by binding to Mϕ via Fcγ receptors and by classical pathway complement activation (Shishido et al., 2012). Therefore, they have also been described as biomarkers for peritoneal injury (Kanda et al., 2013).

Again, rapid *cytokine* and *chemokine* release occurs in peritonitis by innate immune cells from the peritoneal cavity and omentum (Badiu et al., 2011; Coelho et al., 2007). Such glycoproteins promote cell recruitment to the infection site and cell differentiation into effector cells with antimicrobial functions. Therefore, these proteins are crucial in peritonitis and also serve as biomarkers for disease progression.

Antimicrobial peptides (AMP) represent other humoral components with antibacterial properties. AMP can be produced by mesothelial and immune cells, as well as by bacteria. They can anchor to the bacterial membrane and cause damage through pore formation (Mahlapuu et al., 2016). AMPs with no antibacterial capacities but only antiendotoxin and immunomodulatory activities also exist, such as the innate defense regulatory (IDR) or host defense peptides (HDP) (Hilchie et al., 2013; Moretta et al., 2021). Different human HDP can be constitutively expressed in peritoneal cavity (Pottanat et al., 2019). They were initially studied for their antimicrobial activity, but are now known to exhibit more immunomodulatory properties than merely antimicrobial (Nuti et al., 2017). LL-37 is a well-known HDP, also called cathelicidin antimicrobial peptide (CAMP). Cathelicidin enhances phagocytosis by Mϕ and improves sepsis outcome (Boucher et al., 2021; Scheenstra et al., 2020). Defensins are also AMPs present in peritoneal cavity (Zarrinkalam et al., 2001) and new interesting functions have been attributed to them, such as antibiofilm formation (de la Fuente-Núñez et al., 2014).

3. Innate immune response-based therapies in peritoneal bacterial infections

Peritoneal bacterial infections have been long treated with broad spectrum antibiotics. For instance, third generation cephalosporins have been successfully used in SBP as they cover the most important causative agents, such as Enterobacteriaceae. However, growing incidence of nosocomial and non-nosocomial gram-positive and multidrug resistant (MDR) bacterial infections has changed this scenario. There is now a clear need to find new therapeutic strategies for these escalating cases (Maraolo et al., 2018). In this section, innate immune response-based therapies alternative or adjunctive to antibiotics currently under development will be discussed. This will include endogenous molecules with antimicrobial properties, as well as replacement and inhibitory immune system treatments.

In addition to the humoral components mentioned in the previous section, some molecules are also proposed as new therapeutic strategies. For instance, clinical trials using vitamin D, which upregulates cathelicidin levels, have already been used in sepsis patients (Leaf et al., 2014). Similar approaches have been described for soluble proteins with PRR function, such as soluble members of the scavenger receptor cysteine-rich super family (SRCR-SF). One of these SRCR-SF members with bacterial-binding properties is the lymphocyte receptor CD6, in which our group has been extensively working. Specifically, therapeutic properties have been found using a soluble recombinant protein from human CD6 (rshCD6) in mouse models of LPS-induced septic shock, Gram-negative and Gram-positive-induced peritonitis and polymicrobial sepsis caused by CLP (Velasco-de Andrés et al., 2020). More recently, small peptides (9-mer and 11-mer long) mapping to the bacterial binding domains of CD6 have also shown therapeutic properties in mouse models of polymicrobial peritonitis (CLP) (Martínez-Florensa et al., 2018). These results place small proteins and peptides from endogenous origin as alternative and/or adjunctive therapies to antibiotics to treat peritoneal bacterial infections.

Moreover, administration of different cytokines or growth factors has also been tested as immunotherapy. In the case of GM-CSF administration, it has been useful to increase granulocyte populations and in a peritonitis clinical trial showed great results in combination with antibiotic treatment (Chauhan et al., 2018). Treatment with recombinant IFN-γ to patients with sepsis and decreased monocyte function showed reversal of monocyte dysfunction and improved sepsis survival (Nalos et al., 2012). IFN-γ may be more effective in sepsis populations if targeted to patients who are at risk for immune suppression, decreased monocyte function, adaptive immune dysfunction, or chronic inflammation during prolonged hospital stays, and may be even more efficacious if administered in a time-phased approach in conjunction with GM-CSF and G-CSF, or even IL-7 and/or IL-15 cytokines, in order to bolster specific immune function, reduce secondary and nosocomial infections, and improve long-term survival as sepsis recovery evolves.

In addition to administering substitutive or modulating molecules to boost the immune response, there is an option to block proteins responsible of the damage caused by an exacerbated immune response. Administration of monoclonal antibodies against TNF-α prevented sepsis symptoms development (Tracey et al., 1987) with no clinical benefit or little success in experimental models (Lorente and Marshall, 2005). The case of IL-1 inhibitors is very similar to that of TNF-α blockers as some benefits were initially observed but these were unconfirmed in subsequent studies (Fisher et al., 1994; Opal et al., 1997).

4. Conclusions

Peritoneal bacterial infections are a major menace as they often occur in complex clinical contexts such as chronic diseases like cirrhosis. This situation is aggravated by the emergence of multidrug resistant strains, which make it imperative to develop new immune-based therapies alternative or adjunctive to conventional broad-spectrum antibiotics. In this sense, it is critical to know which immune cell populations and mediators are linked to peritoneal infections and which may also serve as biomarkers. Correct management of immune response in peritoneal bacterial infections is crucial to the final disease outcome.

Acknowledgments

F.L. is supported by Spanish Ministerio de Economía y Competitividad (MINECO; PID2019-106658RB-I00) cofinanced by European Development Regional Fund (ERDF) "A way to achieve Europe," and Agència de Gestió d'Ajuts Universitaris i de Recerca (AGAUR; 2017/SGR/1582) from Generalitat de Catalunya. MV-dA, CC, LC-S, SC-L, and AL-P are recipients of fellowships from MINECO (BES-2014-069237; BES-2017-082107; PRE2020-093993), Ministerio de Educación, Cultura y Deporte (FPU15/02897), Chilean Agencia Nacional de Investigación y Desarrollo (ANID; 2018-72190154), respectively.

Disclosures

F.L. is founder and ad-honorem scientific advisor of Sepsia Therapeutics S.L. The rest of authors declare no competing interests.

References

Abt, M.C., et al., 2015. Innate immune defenses mediated by two ILC subsets are critical for protection against acute clostridium difficile infection. Cell Host Microbe 18 (1), 27.

Aithal, G.P., et al., 2021. Guidelines on the Management of Ascites in cirrhosis. Gut 70 (1), 9–29.

Alexander, J.S., Ganta, V.C., Jordan, P.A., Witte, M.H., 2010. Gastrointestinal lymphatics in health and disease. Pathophysiology 17 (4), 315.

Aziz, M., Holodick, N.E., Rothstein, T.L., Wang, P., 2015. The role of B-1 cells in inflammation. Immunol. Res. 63 (1–3), 153–166.

Badiu, D.C., Paunescu, V., Aungurenci, A., Pasarica, D., 2011. Proinflammatory cytokines in peritonitis. J. Med. Life 4 (2), 158.

Bass, G.A., et al., 2021. Tertiary peritonitis: considerations for complex team-based care. Eur. J. Trauma Emerg. Surg. 48 (2), 811–825.

Baumgarth, N., 2011. The double life of a B-1 cell: self-reactivity selects for protective effector functions. Nat. Rev. Immunol. 11 (1), 34–46.

Baumgarth, N., Tung, J.W., Herzenberg, L.A., 2005. Inherent specificities in natural antibodies: a key to immune defense against pathogen invasion. Springer Semin. Immunopathol. 26 (4), 347–362.

Bénézech, C., et al., 2015. Inflammation-induced formation of fat-associated lymphoid clusters. Nat. Immunol. 16 (8), 819–828.

Borregaard, N., 2010. Neutrophils, from marrow to microbes. Immunity 33 (5), 657–670.

Boucher, E., Brown, L., Lahiri, P., Cobo, E.R., 2021. Peritoneal macrophages are impaired in cathelicidin-deficient mice systemically challenged with Escherichia Coli. Cell Tissue Res. 383 (3), 1203–1208.

Brestoff, J.R., et al., 2014. Group 2 innate lymphoid cells promote Beiging of white adipose tissue and limit obesity. Nature 519 (7542), 242–246.

Buonocore, S., et al., 2010. Innate lymphoid cells drive Interleukin-23-dependent innate intestinal pathology. Nature 464 (7293), 1371–1375.

Byrne, P., McGuirk, P., Todryk, S., Mills, K.H.G., 2004. Depletion of NK cells results in disseminating lethal infection with Bordetella pertussis associated with a reduction of antigen-specific Th1 and enhancement of Th2, but not Tr1 cells. Eur. J. Immunol. 34 (9), 2579–2588.

Carrier, P., et al., 2014. L'ascite Non Liée à La Cirrhose: Physiopathologie, Diagnostic et Étiologies. Rev. Med. Interne 35 (6), 365–371.

Chauhan, T.S., et al., 2018. Evaluation of role of granulocyte macrophage Colony stimulating factor (GM-CSF) in treatment of patients with perforation peritonitis: a prospective study. Int. J. Surg. 5 (5), 1902–1907.
Clements, T.W., Tolonen, M., Ball, C.G., Kirkpatrick, A.W., 2021. Secondary peritonitis and intra-abdominal sepsis: an increasingly global disease in search of better systemic therapies. Scand. J. Surg. 110 (2), 139–149.
Coelho, A.L., et al., 2007. The chemokine CCL6 promotes innate immunity via immune cell activation and recruitment. J. Immunol. 179 (8), 5474–5482.
Culligan, K., et al., 2014. A detailed appraisal of Mesocolic Lymphangiology - an immunohistochemical and stereological analysis. J. Anat. 225 (4), 463–472.
Czermak, B.J., et al., 1999. Protective effects of C5a blockade in sepsis. Nat. Med. 5 (7), 788–792.
de la Fuente-Núñez, C., et al., 2014. Broad-Spectrum anti-biofilm peptide that targets a cellular stress response. PLoS Pathog. 10 (5), e1004152.
De Smedt, T., et al., 1996. Regulation of dendritic cell numbers and maturation by lipopolysaccharide in vivo. J. Exp. Med. 184 (4), 1413–1424.
Esin, S., et al., 2008. Direct binding of human NK cell natural cytotoxicity receptor NKp44 to the surfaces of mycobacteria and other bacteria. Infect. Immun. 76 (4), 1719–1727.
Fisher, C.J., et al., 1994. Recombinant human interleukin 1 receptor antagonist in the treatment of patients with sepsis syndrome: results from a randomized, double-blind, placebo-controlled trial. JAMA 271 (23), 1836–1843.
Frydrych, L.M., et al., 2019. GM-CSF Administration improves defects in innate immunity and sepsis survival in obese diabetic mice. J. Immunol. 202 (3), 931–942.
Girardi, G., Lingo, J.J., Fleming, S.D., Regal, J.F., 2020. Essential role of complement in pregnancy: from implantation to parturition and beyond. Front. Immunol. 11, 1681.
Godshall, C.J., et al., 2003. Natural killer cells participate in bacterial clearance during septic peritonitis through interactions with macrophages. Shock (Augusta, Ga.) 19 (2), 144–149.
Gonzaga, R., Matzinger, P., Perez-Diez, A., 2011. Resident peritoneal NK cells. J. Immunol. 187 (12), 6235–6242.
Gordon, S., 2016. Phagocytosis: an Immunobiologic process. Immunity 44 (3), 463–475.
Grönwall, C., Vas, J., Silverman, G.J., 2012. Protective roles of natural IgM antibodies. Front. Immunol. 3, 66.
Guo, R.F., Ward, P.A., 2005. Role of C5a in inflammatory responses. Annu. Rev. Immunol. 23, 821–852.
Ha, S.A., et al., 2006. Regulation of B1 cell migration by signals through toll-like receptors. J. Exp. Med. 203 (11), 2541–2550.
Haidl, I.D., Meghnem, D., Issekutz, T.B., Marshall, J.S., 2020. Toll-like receptor 2 activation induces C-C chemokine receptor 2-dependent natural killer cell recruitment to the peritoneum. Immunol. Cell Biol. 98 (10), 854–867.
Hayakawa, K., Hardy, R.R., Herzenberg, L.A., Herzenberg, L.A., 1985. Progenitors for Ly-1 b cells are distinct from progenitors for other b cells. J. Exp. Med. 161 (6), 1554–1568.
Hilchie, A.L., Wuerth, K., Hancock, R.E.W., 2013. Immune modulation by multifaceted cationic host defense (antimicrobial) peptides. Nat. Chem. Biol. 9 (12), 761–768.
Hotchkiss, R.S., et al., 1999. Overexpression of Bcl-2 in transgenic mice decreases apoptosis and improves survival in sepsis. J. Immunol. 162 (7), 414.
Hsu, H., et al., 2015. WC1 is a hybrid Γδ TCR coreceptor and pattern recognition receptor for pathogenic bacteria. J. Immunol. 194 (5), 2280–2288.
Hültner, L., Blueüunarm, H., Manuel, D., Echutnacher, B., 1996. Critical protective role of mast cell-derived tumor necrosis factor in a model of acute septic peritonitis: involvement of substance P? Exp. Hematol. 381 (6577), 75–77.

Jamur, M.C., et al., 2010. Mast cell repopulation of the peritoneal cavity: contribution of mast cell progenitors versus bone marrow derived committed mast cell precursors. BMC Immunol. 11, 32.

Kanda, R., et al., 2013. Pentraxin 3 as a new biomarker of peritoneal injury in peritoneal dialysis patients. *Journal of.* Artif. Organs 16 (1), 66–73.

Kawahara, T., et al., 2003. Peritoneal cavity B cells are precursors of splenic IgM natural antibody-producing cells. J. Immunol. 171 (10), 5406–5414.

Kelly-Scumpia, K.M., et al., 2011. B cells enhance early innate immune responses during bacterial sepsis. J. Exp. Med. 208 (8), 1673–1682.

Kinjo, Y., et al., 2005. Recognition of bacterial glycosphingolipids by natural killer T cells. Nature 434 (7032), 520–525.

Koulaouzidis, A., et al., 2007. Spontaneous bacterial peritonitis. Postgrad. Med. J. 83 (980), 379–383.

Krensky, A.M., Clayberger, C., 2009. Biology and clinical relevance of Granulysin. Tissue Antigens 73 (3), 193–198.

Lauw, F.N., et al., 2000. Soluble granzymes are released during human endotoxemia and in patients with severe infection due to gram-negative bacteria. J Infect Dis 182 (1), 206–213.

Leaf, D.E., et al., 2014. Randomized controlled trial of calcitriol in severe sepsis. Am. J. Respir. Crit. Care Med. 190 (5), 533–541.

Le-Barillec, K., et al., 2005. Roles for T and NK cells in the innate immune response to shigella Flexneri. J. Immunol. 175 (3), 1735–1740.

Lewis, S., Holmes, C., 1991. Host defense mechanisms in the peritoneal cavity of continuous ambulatory peritoneal dialysis patients. First of two parts. Perit. Dial. Int. 11 (1), 14–21.

Liu, M., Silva-Sanchez, A., Randall, T.D., Meza-Perez, S., 2021. Specialized immune responses in the peritoneal cavity and Omentum. J. Leukoc. Biol. 109 (4), 717–729.

Lorente, J.A., Marshall, J.C., 2005. Neutralization of tumor necrosis factor in preclinical models of sepsis. Shock 24 (SUPPL. 1), 107–119.

Lu, C.-C., et al., 2014. NK cells kill mycobacteria directly by releasing perforin and Granulysin. J. Leukoc. Biol. 96 (6), 1119–1129.

Lynch, L., et al., 2009. Invariant NKT cells and CD1d + cells amass in human Omentum and are depleted in patients with cancer and obesity. Eur. J. Immunol. 39 (7), 1893–1901.

Mahlakõiv, T., et al., 2019. Stromal cells maintain immune cell homeostasis in adipose tissue via production of Interleukin-33. Sci. Immunol. 4 (35), eaax0416.

Mahlapuu, M., Håkansson, J., Ringstad, L., Björn, C., 2016. Antimicrobial peptides: an emerging category of therapeutic agents. Front. Cell. Infect. Microbiol. 6, 194.

Maraolo, A.E., et al., 2018. Current and emerging pharmacotherapy for the treatment of bacterial peritonitis. Expert Opin. Pharmacother. 19 (12), 1317–1325.

Marcenaro, E., et al., 2008. Human NK cells directly recognize mycobacterium Bovis via TLR2 and acquire the ability to kill monocyte-derived DC. Int. Immunol. 20 (9), 1155–1167.

Marques, H.S., et al., 2021. Tertiary peritonitis: a disease that should not be ignored. World J. Clin. Cases 9 (10), 2160–2169.

Martínez-Florensa, M., et al., 2018. Conserved bacterial-binding peptides of the scavenger-like human lymphocyte receptor CD6 protect from mouse experimental sepsis. Front. Immunol. 9, 627.

McCully, M.L., Madrenas, J., 2006. Dendritic cells as arbiters of peritoneal immune responses. Perit. Dial. Int. 26 (1), 8–25.

McSharry, B.P., Gardiner, C.M., 2010. The role of NK cells in bacterial infections. In: Natural Killer Cells: At the Forefront of Modern Immunology, pp. 153–175.

Mebius, R.E., 2009. Lymphoid organs for peritoneal cavity immune response: milky spots. Immunity 30 (5), 670–672.

Meza-Perez, S., Randall, T.D., 2017. Immunological functions of the Omentum. Trends Immunol. 38 (7), 526–536.
Moretta, A., et al., 2021. Antimicrobial peptides: a new Hope in biomedical and pharmaceutical fields. Front. Cell. Infect. Microbiol. 11, 668632.
Nalos, M., et al., 2012. Immune effects of interferon gamma in persistent staphylococcal sepsis. Am. J. Respir. Crit. Care Med. 185 (1), 110–112.
Nauseef, W.M., Borregaard, N., 2014. Neutrophils at work. Nat. Immunol. 15 (7), 602–611.
Neill, D.R., et al., 2010. Nuocytes represent a new innate effector leukocyte that mediates Type-2 immunity. Nature 464 (7293), 1367–1370.
Noh, J.Y., et al., 2020. Toll-like receptors in natural killer cells and their application for immunotherapy. J. Immunol. Res. 2020, 2045860.
Nomura, T., et al., 2002. Essential role of Interleukin-12 (IL-12) and IL-18 for gamma interferon production induced by Listeriolysin O in mouse spleen cells. Infect. Immun. 70 (3), 1049.
Nuti, R., et al., 2017. Antimicrobial peptides: a promising therapeutic strategy in tackling antimicrobial resistance. Curr. Med. Chem. 24 (38), 4303–4314.
Oey, R.C., et al., 2018. Bacterascites: a study of clinical features, microbiological findings, and clinical significance. Liver Int. 38 (12), 2199–2209.
Okabe, Y., Medzhitov, R., 2014. Tissue-specific signals control reversible program of localization and functional polarization of macrophages. Cell 157 (4), 832–844.
Opal, S.M., et al., 1997. Confirmatory Interleukin-1 receptor antagonist trial in severe sepsis: a phase III, randomized, double-blind, placebo-controlled, multicenter trial. Crit. Care Med. 25 (7), 1115–1124.
Pannu, H.K., Oliphant, M., 2015. The subperitoneal space and peritoneal cavity: basic concepts. Abdom. Imaging 40 (7), 2710–2722.
Perez-Shibayama, C., et al., 2018. Fibroblastic reticular cells initiate immune responses in visceral adipose tissues and secure peritoneal immunity. Sci. Immunol. 3 (26), eaar4539.
Perrot, I., et al., 2010. TLR3 and rig-like receptor on myeloid dendritic cells and rig-like receptor on human NK cells are both mandatory for production of IFN-γ in response to double-stranded RNA. J. Immunol. 185 (4), 2080–2088.
Pottanat, N.D., et al., 2019. Analysis of the ribonuclease a superfamily of antimicrobial peptides in patients undergoing chronic peritoneal dialysis. Sci. Rep. 9 (1), 7753.
Qiu, F., et al., 2011. Activation of cytokine-producing and antitumor activities of natural killer cells and macrophages by engagement of toll-like and NOD-like receptors. Innate Immun. 17 (4), 375–387.
Rankin, L.C., et al., 2016. Complementarity and redundancy of IL-22-producing innate lymphoid cells. Nat. Immunol. 17 (2), 179–186.
Ranvier, L., 1874. Du Développement et de l'accroissement Des Vaisseaux Sanguins. Arch. Physiol. Norm. Path. 6, 429–449.
Rittirsch, D., et al., 2008. Functional roles for C5a receptors in sepsis. Nat. Med. 14 (5), 551–557.
Rothstein, T.L., et al., 2013. Human B-1 cells take the stage. Ann. N. Y. Acad. Sci. 1285 (1), 97–114.
Roy, B., et al., 2013. An intrinsic propensity of murine peritoneal B1b cells to switch to IgA in presence of TGF-β and retinoic acid. PLoS One 8 (12), e82121.
Sangisetty, S.L., Miner, T.J., 2012. Malignant ascites: a review of prognostic factors, pathophysiology and therapeutic measures. World J. Gastrointest. Surg. 4 (4), 87–95.
Sawant, K.V., et al., 2021. Neutrophil recruitment by chemokines Cxcl1/KC and Cxcl2/MIP2: role of Cxcr2 activation and glycosaminoglycan interactions. J. Leukoc. Biol. 109 (4), 777–791.

Scheenstra, M.R., et al., 2020. Cathelicidins modulate TLR-activation and inflammation. Front. Immunol. 11, 1137.
Segal, A.W., 2005. How neutrophils kill microbes. Annu. Rev. Immunol. 23, 197–223.
Sehgal, R., Calvin Coffey, J., 2014. Historical development of mesenteric anatomy provides a universally applicable anatomic paradigm for complete/Total Mesocolic excision. Gastroenterol. Rep. 2 (4), 245–250.
Shishido, S.N., et al., 2012. Humoral innate immune response and disease. Clin. Immunol. 144 (2), 142–158.
Small, C.-L., et al., 2008. NK cells play a critical protective role in host defense against acute extracellular staphylococcus aureus bacterial infection in the lung. J. Immunol. 180 (8), 5558–5568.
Song, D.S., 2018. Spontaneous bacterial peritonitis. Korean J. Gastroenterol. 72 (2), 56–63.
Sonnenberg, G.F., et al., 2011. CD4 + lymphoid tissue-inducer cells promote innate immunity in the gut. Immunity 34 (1), 122–134.
Sutherland, R.E., et al., 2008. Mast cell IL-6 improves survival from klebsiella pneumonia and sepsis by enhancing neutrophil killing. J. Immunol. 181 (8), 5598–5605.
Thakurdas, S.M., et al., 2007. The mast cell-restricted tryptase MMCP-6 has a critical Immunoprotective role in bacterial infections. J. Biol. Chem. 282 (29), 20809–20815.
Tracey, K.J., et al., 1987. Anti-Cachectin/TNF monoclonal antibodies prevent septic shock during lethal Bacteraemia. Nature 330 (6149), 662–664.
Tsuchiya, T., et al., 2012. NK1.1+ cells regulate neutrophil migration in mice with Acinetobacter Baumannii pneumonia. Microbiol. Immunol. 56 (2), 107–116.
van Baal, J.O.A.M., et al., 2017. The histophysiology and pathophysiology of the peritoneum. Tissue Cell 49 (1), 95–105.
Velasco-de Andrés, M., et al., 2020. Soluble CD5 and CD6: lymphocytic class I scavenger receptors as immunotherapeutic agents. Cell 9 (12), 2589.
Vivier, E., et al., 2018. Innate lymphoid cells: 10 years on. Cell 174 (5), 1054–1066.
Wang, A.W., et al., 2019. The contribution of the Omentum to the outcome from sepsis: an experimental animal study. Shock 52 (6), 604–611.
Yang, Y., et al., 2007. Division and differentiation of natural antibody-producing cells in mouse spleen. Proceedings of the National Academy of Sciences of the United States of America 104 (11), 4542–4546.
Zarrinkalam, K.H., et al., 2001. Expression of defensin antimicrobial peptides in the peritoneal cavity of patients on peritoneal dialysis. Perit. Dial. Int. 21 (5), 501–508.
Zwicky, S.N., Stroka, D., Zindel, J., 2021. Sterile injury repair and adhesion formation at serosal surfaces. Front. Immunol. 12, 684967.

CHAPTER FOUR

The tumor immune microenvironment in peritoneal carcinomatosis

Jesse Demuytere[a,b,†], Sam Ernst[a,b,c,†], Judith van Ovost[a,b,†], Sarah Cosyns[a,b], and Wim Ceelen[a,b,*]

[a]Experimental Surgery Lab, Department of Human Structure and Repair, Faculty of Medicine and Health Sciences, Ghent University, Ghent, Belgium
[b]Cancer Research Institute Ghent (CRIG), Ghent, Belgium
[c]Laboratory of Experimental Cancer Research, Department of Human Structure and Repair, Faculty of Medicine and Health Sciences, Ghent University, Ghent, Belgium
*Corresponding author: e-mail address: wim.ceelen@ugent.be

Contents

† These authors contributed equally.

International Review of Cell and Molecular Biology, Volume 371
ISSN 1937-6448
https://doi.org/10.1016/bs.ircmb.2022.04.015

Abstract

One in four patients with colorectal cancer, 40% of gastric cancer patients, and 60% of ovarian cancer patients will develop peritoneal metastases (PM) in the course of their disease. The outcome of patients with widespread PM remains poor with currently available treatments. Despite the relatively common occurrence of PM, little is known on the pathophysiology that drives the peritoneal metastatic cascade. It is increasingly recognized that the stromal components of the peritoneal microenvironment play an essential role in tumor progression. However, little is known about the specific interactions and components of the peritoneal tumor microenvironment, particularly with respect the immune cell population. We summarize the current knowledge of the tumor immune microenvironment (TIME) in peritoneal metastases originating from the three most common origins: ovarian, gastric, and colorectal cancer. Clearly, the TIME is highly heterogeneous and its composition and functional activity differ according to tumor type and, within the same patient, according to anatomical location. The TIME in PM remains to be explored in detail, and further elucidation of their immune contexture may allow biology driven design of novel immune modulating or immune targeting therapies.

Abbreviations

CAF cancer-associated fibroblast
CAR T cell chimeric antigen receptor T cell
CCL chemokine (C—C motif) ligand
CCR C—C chemokine receptor
CD cluster of differentiation
cDC conventional dendritic cell
CEA carcinoembryonic antigen
CLR C-type lectin receptor
COX-2 cyclooxygenase-2
CRC colorectal cancer
CT zone central tumoral zone
CTL cytotoxic T lymphocyte
CTLA4 cytotoxic T lymphocyte-associated antigen 4
CXCL chemokine (C—X—C motif) ligand
CXCR C—X—C chemokine receptors
DC dendritic cell
ECM extracellular matrix
EGF epidermal growth factor
EPA extravasated platelet aggregation
EpCAM epithelial cell adhesion molecule
FALC fat-associated lymphoid cluster
FGF fibroblast growth factor
FOXP3 forkhead box P3
FRC fibroblastic reticular cell
GATA6 GATA binding protein 6
GC gastric cancer
HIPEC hyperthermic intraperitoneal chemotherapy
HLA-DR human leukocyte antigen DR
HOXA9 homeobox A9

IF	immunofluorescence
IFNγ	interferon γ
Ig	immunoglobulin
IGF2R	insulin-like growth factor 2 receptor
IHC	immunohistochemistry
IL	interleukin
ILC	innate lymphoid cell
IM	invasive margin
IRF4	interferon regulatory factor 4
KIR	killer-cell immunoglobulin-like receptor
LAG3	lymphocyte activation gene 3
LPM	large peritoneal macrophage
MC	mesothelial cell
MCP-1	monocyte chemoattractant protein 1
MDSC	myeloid-derived suppressor cell
MHC	major histocompatibility complex
MMP	matrix metalloproteinase
MMT	mesothelial-to-mesenchymal transition
NET	neutrophil extracellular trap
NK cell	natural killer cell
NKT cell	natural killer T cell
OC	ovarian cancer
OS	overall survival
pCRC	primary colorectal cancer
pDC	plasmacytoid dendritic cell
PDGF	platelet-derived growth factor
PD-1	programmed cell death protein 1
PD-L1	PD-1 ligand 1
PEL	peritoneal elastic lamina
PF	peritoneal fluid
pGC	primary gastric cancer
PM	peritoneal metastases
pOC	primary ovarian cancer
PRG2	proteoglycan 2
RIG	retinoic acid-inducible gene
RLR	RIG-I-like receptor
RT-PCR	reverse transcription polymerase chain reaction
RUNX1	runt-related transcription factor 1
scRNA-seq	single-cell RNA sequencing
SPM	small peritoneal macrophage
STING	stimulator of interferon genes
TAM	tumor-associated macrophage
TGF-β	transforming growth factor β
T_{fh} cell	follicular helper T cell
T_h cell	T helper cell
TIL	tumor-infiltrating lymphocyte
TIME	tumor immune microenvironment
TIM3	T cell membrane protein 3

TLR	toll-like receptor
TME	tumor microenvironment
TNF	tumor necrosis factor
Treg	regulatory T cell
VEGF	vascular endothelial growth factor
VISTA	V-domain Ig suppressor of T cell activation
WT1	Wilms' tumor 1

1. Introduction to the normal peritoneal microenvironment

1.1 Composition of the peritoneal microenvironment

The peritoneal cavity contains a small volume (approx. 10–100 mL in physiological conditions) of peritoneal fluid (PF), which contains growth factors, nutrients, cytokines, and chemokines as well as leukocytes, all of which are continuously exchanged between the PF and the blood. Recent cadaveric research showed that the mean peritoneal surface area in the adult female is approximately 1.43 m^2, with the small bowel and its mesentery accounting for 39% of the total surface (Albanese et al., 2009). Both the interior aspect of the abdominal wall and the viscera are covered by a peritoneal layer consisting of mesothelial cells (MCs), which are flat, squamous like cells approximately 25 μm in diameter, and by an underlying extracellular matrix (ECM) (Mutsaers et al., 2016). The MCs provide a protective, non-adhesive surface by production of surfactant, proteoglycans, and glycosaminoglycans. Also, they have a role in transport of fluid and cells across the peritoneal cavity, present antigens to T cells, and participate in inflammation and tissue repair by the secretion of chemokines (MCP-1 and IL-8), cytokines (IL-6 and IL-1), growth factors (TGF-β, FGF, PDGF), ECM components, and other biological mediators (Yung et al., 2006). Mesothelial cells display pattern-recognition receptors, such as Toll-like receptors (TLRs), RIG-I–like receptors (RLRs), and C-type lectin receptors (CLRs), which recognize microbial pathogens and stimulate release of inflammatory mediators and immune pathways (Mutsaers et al., 2016). The MCs are in constant communication with the cellular and molecular constituents of the peritoneal fluid, and both constitute a peritoneal ecosystem.

The lymphatic system of the peritoneal cavity is comprised of submesothelial lymphatics, the network of lymph nodes and channels, and specialized structures including lymphatic lacunae and the omentum.

Yonemura reported the concept of "translymphatic" metastasis, allowing direct migration and absorption of fluid, particulate matter, and cancer cells into the lymphatic circulation (Yonemura et al., 2020). This route of systemic absorption occurs at specialized anatomical and functional structures, including diaphragmatic stomata and milky spots. Stomata are 8–10 μm wide fenestrations between the MCs which provide direct access to an underlying network of lymphatic vessels, lacunae, and cisternae that allows to clear particulate matter, bacteria, and cells from the peritoneal cavity into the systemic circulation (Michailova et al., 2005; Ohtani et al., 2001).

Milky spots or fat-associated lymphoid clusters (FALCs) are dedicated immunological niches consisting of glomerulus-like conglomerates of capillaries, immune cells, and fibroblastic stromal cells covered by a fenestrated layer of cuboid MCs (Fig. 1). In humans, the diameter of milky spots ranges from 350 to 800 μm, and their density seems to decrease with age from around 50 to 10 per cm^2 (Krist et al., 1997; Liu et al., 2016; Mutsaers et al., 2016; Shimotsuma et al., 1991). Recent data suggest that FALCs are covered by specialized $CXCL1^+$ mesothelial cover cells, which coordinate recruitment of neutrophils and their subsequent expression of neutrophil extracellular traps (NET) in inflammatory conditions (Jackson-Jones et al., 2020). FALCs are populated by a specific, CCL19 expressing type of fibroblastic reticular cells (FRCs) which orchestrate immune cell recruitment and FALC remodeling (Perez-Shibayama et al., 2018). In addition to B cells, macrophages, T cells, dendritic cells (DCs), and innate lymphoid cells (ILCs) have been identified within FALCs. However, unlike conventional lymphoid organs, T cells cluster around blood vessels, instead of within a defined T cell zone (Trim and Lynch, 2021).

Despite its role in defense against bacterial infection and the presence of immune cells in FALCs, the omentum is recognized as an organ at risk for peritoneal cancer dissemination. The biological processes underlying this tropism are not fully understood. Mechanisms that have been explored include NET formation and mechanical trapping of cancer cells, reprogramming of tumor metabolism by omental adipocytes, and pro-tumorigenic signaling by host cells such as omental adipocytes, MCs, and fibroblasts (Ladanyi et al., 2018; Lee et al., 2019; Sommerfeld et al., 2021).

1.2 Innate immunity

In mice, two main types of resident peritoneal macrophages were identified: GATA6-dependent large peritoneal macrophages (LPMs), and monocyte-derived, IRF4-dependent small peritoneal macrophages (SPMs)

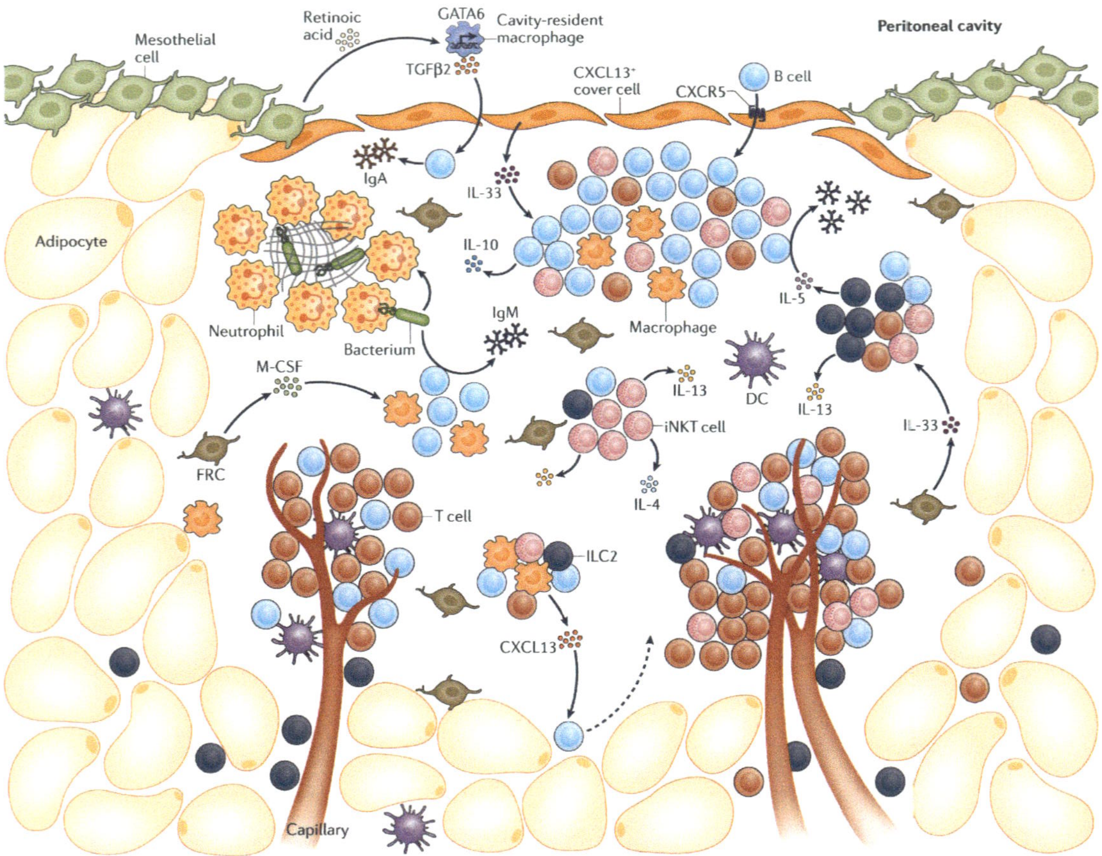

Fig. 1 Immune environment in the healthy peritoneal cavity. Fat-associated lymphoid structures (FALCs) are dense immunological structures found within the omentum. FALCs are capped by a layer of CXCL13 + FALC cover cells interspersed with the mesothelial cells. FALCs also contain a network of fibroblastic reticular cells (FRCs) which recruit and modulate immune cells. Mesothelial cells also release retinoic acid, which sustains GATA6-expressing cavity-resident macrophages in the peritoneum, which induce B cell class-switching to IgA via transforming growth factor-β2 (TGFβ2). Neutrophils also home to FALCs during acute inflammation and form neutrophil extracellular traps that capture contaminants or cancer cells. *Reprinted with permission from Trim, W.V., Lynch, L., 2021. Immune and non-immune functions of adipose tissue leukocytes. Nat. Rev. Immunol. 6. doi:10.1038/S41577-021-00635-7.*

(Kim et al., 2016; Rosas et al., 2014). GATA6 expression by LPMs is induced by the vitamin A derivative retinoic acid (RA), the conversion of which is controlled by retinaldehyde dehydrogenase 1 and 2, both highly expressed by MCs and omental stromal cells that express the transcription factor Wilms' Tumor 1 (WT1) (Buechler et al., 2019). In addition to these fluid borne macrophages, additional distinct macrophage populations were recently identified that reside within the mesenteric and parietal mesothelial linings of the peritoneum, and which bear a striking resemblance with tumor associated macrophages in ovarian cancer (Zhang et al., 2021). The presence of granulocytes in the peritoneal cavity is stimulated by infection or inflammation, which triggers the expression of chemokines and cytokines by MCs and omental stromal cells. DCs are a heterogeneous population of innate immune cells generally classified as conventional cDC1, cDC2, monocyte-derived DC, and plasmacytoid DC (pDC). In mice, cDC2 represent the predominant population in the omentum (Carlow et al., 2009). Little is known on the presence and phenotype of DCs in the healthy human peritoneal cavity. In peritoneal effluent from dialysis patients, the DC population was identified as cDC2 (Liao et al., 2017).

1.3 Adaptive immunity

B cells play an important role in the peritoneal immunity. In mice, the peritoneal cavity is mainly populated by B1 cells, which display distinct developmental, phenotypic, and functional characteristics compared to conventional B2 cells, a small number of which is found in the peritoneal cavity and omentum (Haas, 2015; Hastings et al., 2006). The number of peritoneal B lymphocytes in humans seems to be much lower compared to mice (Nisitani et al., 1997). However, the phenotypical and functional characteristics of peritoneal B cells in humans remain to be elucidated. Analysis of human omental samples showed only very scarce immune cell populations, consisting mainly of quiescent, isolated, or clustered B cells and isolated plasma cells, the IgG isotype being the more abundant one (Boursier et al., 2006). Similarly, only sparse data are available on T cell populations in the healthy peritoneal cavity and omentum. Wu and coworkers found that omental regulatory T cells (Tregs) adopt a tissue specific phenotype, different from thymic Tregs (Wu et al., 2019).

2. The peritoneum in metastasis

As described above, the peritoneum possesses several unique anatomic and microenvironmental factors which render it an organ of choice for metastasis of several abdominal malignancies. One in four patients with colorectal cancer, 40% of gastric cancer patients, and 60% of all ovarian cancer patients will develop peritoneal metastases in their disease course (Burg et al., 2020; Ceelen et al., 2020; Sugarbaker, 2018).

The process of peritoneal metastasis is a continuous and stepwise process, wherein cancer cells detach from their primary location, are transported by the peritoneal fluid, and attach to the peritoneum, followed by invasion into the subperitoneal space, proliferation, and angiogenesis. However, different cancer types possess unique mechanisms by which these steps are accomplished (Lemoine et al., 2016; Mikuła-Pietrasik et al., 2018). Following Paget's "seed and soil" hypothesis, the concept of a peritoneal pre-metastatic niche, and the mechanisms by which the peritoneum contributes to metastasis are becoming clearer. Whereas it was previously assumed that the peritoneum exerted an anti-tumoral function, or at most was an innocent bystander in metastasis, recent insights have implicated the peritoneal microenvironment as an active contributor in forming a favorable pre-metastatic niche in response to tumor stimuli. Both macrophages and MCs, the main cellular components of the peritoneal microenvironment, can exert pro-metastatic effects (Mikuła-Pietrasik et al., 2018). The presence of an increased number of macrophages in a distant biopsy of the peritoneum at the time of surgery for primary gastric cancer was significantly associated with eventual peritoneal relapse (Song et al., 2019).

Furthermore, next to production of pro-inflammatory cytokines and ECM components, MCs can undergo phenotypic plasticity in response to inflammatory stimuli, acquiring a mesenchymal phenotype in a process termed mesothelial-to-mesenchymal transition (MMT). First described in the context of peritoneal dialysis, this process has been observed in peritoneal metastases originating from different primary cancers and is an important contributor to the recruitment of cancer-associated fibroblasts (CAFs) (Demuytere et al., 2020; Rynne-Vidal et al., 2015).

Next to these factors, research into how the microenvironment of the peritoneum shapes metastatic potential is made attractive due to both the high unmet clinical need in treating peritoneal metastases, and the relative ease by which new therapies can be introduced, owing to the anatomical location and the possibility of local (intraperitoneal) treatment which allows to diminish systemic and off-target effects (Ceelen et al., 2020).

3. The tumor immune microenvironment

The tumor microenvironment (TME) of solid cancers not only comprises tumor cells but also vascular endothelial cells, fibroblasts, and a variety of innate and adaptive immune cells, as well as structural (ECM) components and an array of extracellular soluble molecules (cytokines, chemotactic factors, growth factors, etc.) (Duan et al., 2020). Cancer cells create and educate the TME to aid in the establishment of cancer hallmarks, respond to external or internal stress, stimulation or treatment, and eventually aid in the survival and expansion of cancer cells (Jin and Jin, 2020). Cancer biology and development cannot be understood solely via the features of cancer cells; the TME's contributions to carcinogenesis must also be considered (Duan et al., 2020). In this section an introduction to the immune component of the TME is discussed.

3.1 Composition

The tumor immune microenvironment (TIME) is a complex ecosystem containing adaptive and innate immune cells with tumor-promoting and anti-tumoral roles (Gajewski et al., 2013). New technologies such as single-cell RNA sequencing (scRNA-seq) and mass cytometry have recently been developed and applied in tumor immunology to help delineate the landscape within the TME at the single-cell level, making it easier to observe and identify new subsets of tumor-associated immune cells (Lei et al., 2020). Immune cells in the TME include immunosuppressive cells such as tumor-associated macrophages (TAMs), myeloid-derived suppressor cells (MDSCs), and Tregs, as well as anti-tumor effector cells such as cytotoxic $CD8^+$ T cells, $CD4^+$ T helper (T_h) 1 cells, B cells, mast cells, granulocytes, DCs, and natural killer (NK) cells (Duan et al., 2020; Jin and Jin, 2020). Immune cells in the TIME are still being characterized regarding their diversity, flexibility, and activities, as well as their involvement in influencing the response to immunotherapies (Duan et al., 2020).

The composition of immune infiltrates and their functional state within the TME of equivalent tumors can vary greatly between patients, implying that mapping these characteristics within the TME is critical for both diagnosis and treatment strategy design. There have been attempts to categorize the immune responses in the TIME either as "cold" (non-T cell inflamed) or as “hot” (T cell inflamed), where the levels of proinflammatory cytokine production and T cell infiltration have a significant role (Duan et al., 2020). However, other classification systems such as the Immunoscore also focus on location of the infiltrating immune cells. These cells are not randomly dispersed, but rather organized in more or less dense infiltrates in the central tumoral zone (CT), the invasive margin (IM) of tumoral nests, and lymphoid islets next to the tumor (Galon et al., 2012). Later on, the altered phenotype was added to the binary hot/cold classification and further separated into "excluded" and "immunosuppressed." T cells have been discovered in the IM without being able to enter the center of the tumor in some situations. This “excluded” phenotype reflects the host immune system’s inherent ability to produce an efficient T cell-mediated immunological response, as well as the tumor’s ability to evade such a response by physically impeding T cell invasion (Galon and Bruni, 2019). However, in the “immunosuppressed” phenotype, tumor sites have a low level of immune infiltration in the CT as well as in the IM resulting in the absence of a coordinated immune response. The immunosuppressive environment restricts further recruitment and expansion. This limited immune infiltration and expansion has also been described as an “immune desert” (Galon and Bruni, 2019). A high density of T_h1 or cytotoxic memory T lymphocytes in the initial CT and IM is linked to extended disease-free and overall survival, as well as a low probability of relapse and metastasis in human cancers. This was demonstrated in colorectal cancer and may apply to the majority of human cancers. The identification of these immune phenotypes could aid in the development of predictive biomarkers for immunotherapeutic classes (Gajewski et al., 2013). The fact that some patients develop an extensive immune response while others do not, or not sufficiently, poses the fundamental question of the molecular mechanisms underlying these phenotypes. Sources of interpatient heterogeneity in individual cancer patients include host differences in tumor cells as well as germ-line mutations and environmental influences (Gajewski et al., 2013).

3.2 Function

In the TME, immune cells play a significant role in carcinogenesis. These tumor-associated immune cells may have tumor-antagonizing ($CD8^+$ cytotoxic T cells and effector $CD4^+$ T cells, NK cells, DCs, M1-like-polarized macrophages, and neutrophils) or tumor-promoting effects

(Tregs, TAMs, and MDSCs) (Duan et al., 2020; Gajewski et al., 2013; Jin and Jin, 2020; Lei et al., 2020). Although tumor-antagonizing immune cells in the TME tend to target and kill cancer cells early in the carcinogenesis process, cancer cells eventually escape immune surveillance and even suppress the cytotoxic function of these tumor-antagonizing immune cells through a variety of methods (Lei et al., 2020). However, in order to develop new treatments, it is fundamental to define the tumor-antagonizing effects employed by the immune cells of the TME.

Natural killer cells are innate lymphocytes that possess activating receptors such as NKG2D and natural cytotoxicity receptors, which detect stress-induced ligands expressed by the majority of tumors. They also express major histocompatibility complex class I (MHC I)-specific inhibitory receptors (KIR and NKG2A), which preferentially kill MHC I-deficient tumor cells (Li et al., 2021).

$CD8^+$ cytotoxic T lymphocytes (CTLs) have long been thought to be the most important lymphocyte subset for killing cancer cells that present MHC-I markers. Furthermore, the $CD4^+$ T cells can aid in CTL priming and proliferation. The $CD4^+$ subset can either help directly or through activation of DCs (Lei et al., 2020). The CTLs require two signals, one from the T-cell receptor and the other one from immune checkpoint receptors which can be inhibitory or stimulatory. Inhibitory receptors are extensively studied due to their potential as therapeutic target. Some of these targeted receptors and ligands include: programmed cell death protein 1 (PD-1), PD-1 ligand (PD-L1), cytotoxic T-lymphocyte-associated antigen 4 (CTLA4), T cell membrane protein 3 (TIM3), and lymphocyte activation gene 3 (LAG3) (Pardoll, 2012). The inhibitory capacity of these targets is used by cancer cells to dysregulate T cell activity and results in immune resistance (Pardoll, 2012). In addition, the chronic presence of antigens and inflammation within the TME can result in "exhausted" CTLs, contributing to T cell malfunction and tumor growth (Lei et al., 2020). The exhausted phenotype is characterized by overexpressed inhibitory receptors, decreased effector cytokine production and cytolytic activity, leading to the loss of the ability to eliminate cancer (Jiang et al., 2015).

Tumor cells employ a variety of tactics to elude immune surveillance, and the highly immunosuppressive and metabolically stressed TME appears to be an important obstacle to effective antitumor immunity (Duan et al., 2020). Emerging treatment techniques to deplete or diminish tumor-promoting immune cells in the TME, such as Tregs, TAMs, and MDSCs, are being developed (Duan et al., 2020; Jin and Jin, 2020).

Tregs are important in maintaining immunological homeostasis and peripheral tolerance. Their suppressive activity within the TME may prevent CTLs from effectively attacking cancer cells. Indeed, recent scRNA-seq data revealed that Tregs in tumors are highly heterogeneous and clonally expanded, and that Treg function is closely linked to patient prognosis (Lei et al., 2020).

MDSCs are a heterogeneous population of monocytic and polymorphonuclear immature myeloid cells that develop in response to inflammation-induced emergency myelopoiesis. MDSC proliferation hinders the formation of protective anti-tumor immunity in most cancer patients, enhances tumor cell growth and metastasis, and decreases the effectiveness of cancer immunotherapy (Li et al., 2021).

TAMs are one of the most abundant types of cells in the TME, and they display a variety of phenotypes and activities in response to numerous microenvironmental signals provided by tumor and stromal cells. The M1-like polarized macrophage is one of the phenotypes displayed which has pro-tumorigenic effects. Another phenotype frequently found as TAM is the M2-like polarized macrophage. However, due to their plasticity, the macrophages are a heterogenous population that cannot be classified in two categories. The markers defining M1- and M2-like phenotypes can be displayed on many subtypes in different expression levels causing a heterogenous group of both anti-and pro-tumorigenic macrophages (Boutilier and Elsawa, 2021). TME-induced macrophage polarization should be addressed in detail in order to gain a better understanding of TAM-related anti- and pro-tumor effects and the development of new therapeutic methods. Numerous studies have now shown that TAM density and location are linked to poor clinical outcomes in a variety of solid malignancies, including bladder, breast, renal, prostate, and gastric cancer. TAMs are thought to play a crucial role in tumor metastasis through controlling cancer cell EMT, according to recent research (Wei et al., 2019).

Taken together, the available data suggest that the TIME is a heterogenous component which is implicated in numerous processes. Besides describing the function of each individual cell type, it is important to define the mechanisms of action employed by the TIME to influence tumor progression and metastasis. The reciprocal interactions between components of the TME, and their effect on the TIME, remain to be further clarified.

Interaction between tumor cells and immune cells in various tissue environments may lead to development of metastatic tumors that have significantly different TIME composition compared to the primary tumor, which could result in distinct immunotherapy responses (Duan et al., 2020; Zhu et al., 2019). Indeed, a recent study examined immune infiltrates in initial breast tumors and patient-matched metastatic tumors in the brain or ovary, and discovered a substantial decrease in the presence of multiple immune cells, including $CD8^+$ T cells, Tregs, and DCs (Zhu et al., 2019).

Additionally, both the TIME of the primary tumor and those of metastatic sites influence progression: treatment failure in advanced patients with metastasis has been linked to differences in immune microenvironments between primary lesions and metastatic foci (Jin and Jin, 2020).

While direct interactions between tumor cells and immune cells are clearly important, it is also worth noting that immune cell interactions with other TME components might influence tumor fate. Within the TME, there is a complicated sequence of interactions between immune cell types and non-tumor cells that significantly influence tumor development, invasion, and metastasis (Hinshaw and Shevde, 2019). Immune cells interact with stromal cells as well, such as CAFs. Due to their role in processes including ECM-modulation and inflammation, CAFs have been linked to tumor growth, invasion, and metastasis(Gunaydin, 2021).

As a result, not only should tumor—immune cell cross-talk and tumor—stromal cross-talk be considered in therapeutic designs, but also stromal—immune cell cross-talk, as it plays a crucial role in tumor formation (Hinshaw and Shevde, 2019).

4. The TIME in peritoneal metastases

Due to both its unique anatomic relationship with several abdominal organs and microenvironmental factors, the peritoneum is a preferred metastatic site for many cancers. We will specifically highlight the unique properties of peritoneal metastases (PM) arising from the three most common origins: colorectal, gastric, and ovarian malignancies.

4.1 The TIME landscape of colorectal PM

Colorectal cancer (CRC) is the third most common cause of cancer-related death worldwide, and has a 5-year survival rate of 65% (Zhao et al., 2021). Histologically, the most common types are conventional adenocarcinoma, mucinous carcinoma, and signet-ring cell carcinoma (Lemoine et al., 2016). One in four CRC patients develops peritoneal metastases, which are associated with a poor prognosis due to limited effective treatment options

(Ceelen et al., 2020). The TIME in colorectal PM remains incompletely characterized. Several small scale studies were performed on either tissue or malignant ascites from patients with colorectal PM.

Seebauer et al. compared the TIME between primary CRC and colorectal peritoneal metastases using tissue from two different patient groups, and found several differences (Fig. 2). First, compared to primary CRC, tumor cells in PM showed increased senescence (evidenced by H3K9me3, p21Cip1, and CDKN2A) as well as decreased proliferation (evidenced by Ki-67, PCNA, and cyclin D1) (Seebauer et al., 2016).

Second, striking differences were found in the immune contexture: the primary tumors showed significantly increased numbers of Natural Killer (NK) cells, follicular helper T cells (TFH) and B cells, whereas PM showed less $CD4^+$ T cells, $CD8^+$ T cells, eosinophilic granulocytes, T_h17 and Tregs. Also, PM were characterized by significantly increased interferon-γ (IFNγ), upregulation of tumor necrosis factor (TNF), and increased levels of the NK cell-regulating cytokine IL-15. Upregulation of angiogenesis-related genes, including vascular endothelial growth factor-A (VEGF-A), correlated with pronounced neovascularization in PM samples. In addition, peritoneal metastases showed an increased $CD20^+$ B cell population, and slightly higher $CD79a^+$ and $CD138^+$ plasma cells.

Zhao and colleagues performed a similar comparison based on clinical samples, and also found a lower density of $CD3^+$ and $CD8^+$ T cells in PM. Furthermore, they observed that if TILs were present, they were more likely to be situated in the IM rather than in the CT zone. However, in contrast to the findings of Seebauer and colleagues, a lower number of $CD20^+$ cells was found in PM, and these were more often located in the IM (Zhao et al., 2021).

Pesce et al. analyzed the features of $CD56^{dim}$ NK tumor associated cells isolated from peritoneal fluid in patients undergoing surgery for PM. The authors found significant differences between low grade and high grade PM. In the patient with colorectal PM, NK cells were predominantly mature, but showed impaired degranulation and enhanced PD-1 expression (Pesce et al., 2019).

Miller and coworkers isolated plasmacytic dendritic cells (pDCs) in peritoneal fluid from patients with PM of various origins. They found that the pDC population ranged from 0.01% to 4.73%, and this percentage was higher in pancreatic compared to gastrointestinal tract cancer (2.6 vs 0.7%; $P=0.01$) (Miller et al., 2021). Interestingly, neoadjuvant chemotherapy seemed to adversely affect the percentage of pDCs, which were noted to

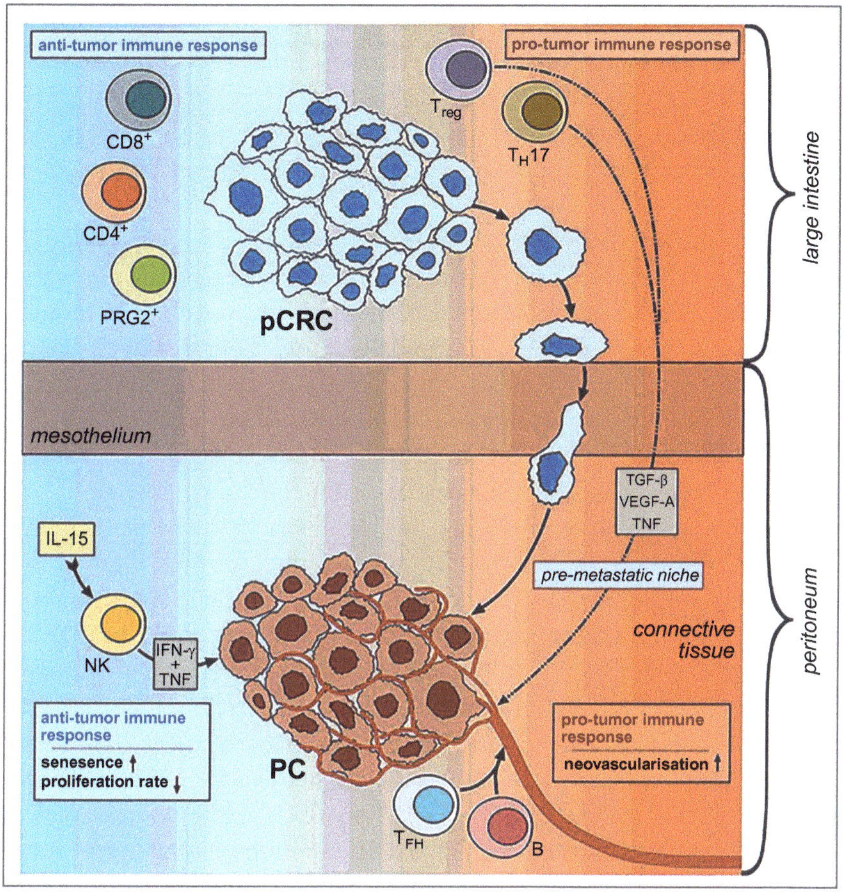

Fig. 2 Comparison of the tumor immune microenvironment of primary colorectal cancer (pCRC) vs colorectal peritoneal carcinomatosis (PC). In the microenvironment of PC, NK cells exert antitumor immune effects, while T_{fh} and B cells may promote tumor growth by inducing neovascularization. The premetastatic niche is supported by Tregs and T_h17 cells from the primary tumor. *Reprinted with permission from Seebauer, C.T., Brunner, S., Glockzin, G., Piso, P., Ruemmele, P., Schlitt, H.J., et al., 2016. Peritoneal carcinomatosis of colorectal cancer is characterized by structural and functional reorganization of the tumor microenvironment inducing senescence and proliferation arrest in cancer cells. OncoImmunology 5, 1. doi:10.1080/2162402X.2016.1242543/SUPPL_FILE/KONI_A_1242543_SM3568.ZIP.*

drop in one patient from 2.05% to 0.58% after a single dose of FOLFOX chemotherapy.

Taken together, the limited available data suggest important differences in the immune contexture between primary CRC and the associated peritoneal metastatic disease.

4.2 The TIME landscape of gastric PM

Forty percent of gastric cancer (GC) patients will develop peritoneal metastases in their disease course, dramatically lowering 5-year survival rates to $<5\%$ (Sugarbaker, 2018). Although common, little research has been done into the specific peritoneal microenvironment compared to that of the primary tumor.

Peritoneal metastases from GC are characterized by severe stromal fibrosis as a result of the interplay between the MMT-process induced by cancer cells, resulting in accumulation of CAFs, and release of fibrosis-inducing cytokines and growth factors such as IL-17A and TGF-β. Gunjigake and colleagues recently reported in an immunohistochemical study of 70 patient samples that the main producers of IL-17A in the peritoneal GC TME are mast cells, while no T_h17 cells, which are canonically important producers of IL-17A, were observed. Furthermore, the functional role of IL-17A secretion was elucidated, showing that MCs are induced to acquire a pro-fibrotic phenotype by IL-17A stimulation. In turn, this fibrosis increased intratumoral pressure, and might act as a barrier to effective tumor cell killing by encapsulating tumor implants (Gunjigake et al., 2021).

TILs are key factors in GC PM. Yamaguchi et al. performed an analysis of the presence of TILs and found T_h cells, cytotoxic T cells, Tregs and MDSCs in clinical samples. Both higher infiltration of $CD8^+$ T cells and an increased ratio of $CD8^+$ T cells vs MDSCs (defined as $CD33^+$ cells) were significantly associated with OS in this patient cohort, demonstrating the key role of effective tumor cell killing by cytotoxic T cells (Yamaguchi et al., 2021). Other TIL subtypes, such as Tregs, have not yet been shown to influence survival in the PM context (Yamaguchi et al., 2020). In contrast to primary GC (pGC), a study on paired primary and PM samples reported lower infiltration of $CD8^+$ T cells in PM, with no differences in $CD4^+$ T cell infiltration (Fujimori et al., 2020). Presence of MDSCs in tumor stroma is positively correlated with increased extravasated platelet aggregation (EPA). Platelets contain large amounts of growth factors, including TGF-β, and might as such influence MDSC accumulation in PM, as EPA has recently been confirmed to be present in more than 50% of PM samples (Yamaguchi et al., 2020).

Transcriptomic techniques can differentiate immune signatures, and dissect the microenvironment in GC PM. An increased amount of resting memory $CD4^+$ T cells were associated with a more aggressive phenotype. An immune profiling based on whole-exome and whole transcriptome sequencing of 44 PM specimens revealed two main groups: the T cell "exclusive" and T cell "exhausted" phenotype. In the exclusive subtype, low levels of cytotoxic T cell markers, checkpoint inhibitors and other immune-related genes were found, pointing toward a subgroup of patients with "immune desertification" in the PM microenvironment. The exhausted subtype is marked by high levels of immune-related genes, including cytotoxic T cell markers such as CD8 and macrophage markers CD68 and CD163, and by increased expression of the checkpoint markers TIM3 and its ligand galectin-9, VISTA, and TGF-β1. Interestingly, no other known checkpoint markers such as PD-1, PD-L1, or CTLA4 were found to be expressed highly in this subgroup, pointing towards a potential therapeutic target (Wang et al., 2020).

In addition to TILs, there is considerable evidence pointing toward the importance of macrophages in the TME of GC PM, both in supporting tumor growth, and in priming of the "pre-metastatic niche." This concept, described in part III of this review, is of considerable importance in the formation of GC PM: the abundance of $CD68^+$ macrophages in a standardized biopsy of the peritoneum at a distance of the primary tumor is a potent predictor of overall survival in GC patients. Interestingly, when comparing groups of patients with PM vs no PM, no differences in the amount of TAMs in the GC tumor bed were observed, indicating that peritoneal macrophages, rather than TAMs in the primary tumor, are a driver of PM (Song et al., 2019).

TAMs are abundant in PM. Comparing pGC and paired PM, Fujimori and colleagues demonstrated significantly higher numbers of $CD163^+$ macrophages in PM (Fujimori et al., 2020), and a comparison between pGC and PM using flow cytometry demonstrated increased total macrophage numbers in PM (Yamaguchi et al., 2016). Most authors define TAMs in the peritoneal context as having an M2-like, or alternatively activated, phenotype (Fujimori et al., 2020; Song et al., 2019; Yamaguchi et al., 2016). Functionally, TAMs isolated from PM of GC exhibit increased ratios of macrophages positive for M2-like markers. Furthermore, RT-PCR analysis confirmed increased expression of M2-like gene expression markers such as IL-10A, VEGF-A, VEGF-C, matrix metalloproteinase (MMP)-1 and amphiregulin (Yamaguchi et al., 2016). Additionally, an increased ratio of macrophages ($CD68^+$) displayed VEGF and epidermal growth factor (EGF)-positivity when comparing peritoneal lavage of GC patients with and without PM, pointing toward functional changes in the PM microenvironment (Song et al., 2019). The increased expression of VEGF observed in TAMs is of particular interest in the formation of PM. In a study by

Song et.al., overexpression of progenitor endothelial cells ($CD34^+$) in the surgical margin of the pGC and the normal peritoneum was significantly increased in patients with PM, pointing to angiogenesis driven by TAMs as an important factor in the development of PM (Song et al., 2019).

To conclude, PM from GC influence the "pre-metastatic niche" of the peritoneum to become more amenable to metastasis, both by fibrosis and activation and recruitment of CAFs and TAMs. PM are marked by increased infiltration of cytotoxic T cells, and an abundance of TAMs. Transcriptomic analyses point toward a marked heterogeneity however, with both an exhausted phenotype and immune desertification being described in this patient cohort. Furthermore, the role of the innate immune system and B cells remain to be defined (Fig. 3).

4.3 The TIME landscape of ovarian PM

Ovarian cancer (OC) is the seventh most common cancer and the eight most common cancer death in women, accounting for 4.3% of all cancer-related deaths (Torre et al., 2018). The most common type, epithelial OC, can be histologically subdivided in high-grade and low-grade serous OC, clear cell OC, endometrioid OC, and mucinous OC. Although early-stage OC has a good prognosis, most women present with late-stage disease at primary diagnosis (Torre et al., 2018). FIGO Stage III and IV OC is associated with the presence of malignant ascites and metastatic spread to the peritoneum. This ascites generally contains cytokines, chemokines, and a cellular fraction, such as cancer cells, immune cells and MCs, and plays an indisputable role in the metastatic process and clinical path of OC patients. The presence of PM is associated with high relapse rates, and thus a bad prognosis. Using the current standard of care the median recurrence-free survival is only 14.2 months with a median OS of $\pm$ 45 months (van Driel et al., 2018). Despite the high prevalence and bad prognosis of PM, the immune compartment of ovarian cancer PM has not received much attention thus far.

The metastatic process in OC results from an interplay between ascites, the peritoneum, and the primary tumor. Wang and colleagues compared the peritoneum and subperitoneal stroma from the pelvic area between patients with malignant vs benign ovarian tumors. They showed that the presence of primary OC (pOC) prepares the normal peritoneum as a premetastatic niche by recruiting immunosuppressive immune cells (Wang et al., 2005). Using bulk RNAseq, they observed an upregulation of CXCL6, CXCL8 (chemoattractants for neutrophils and monocytes/macrophages, respectively) and

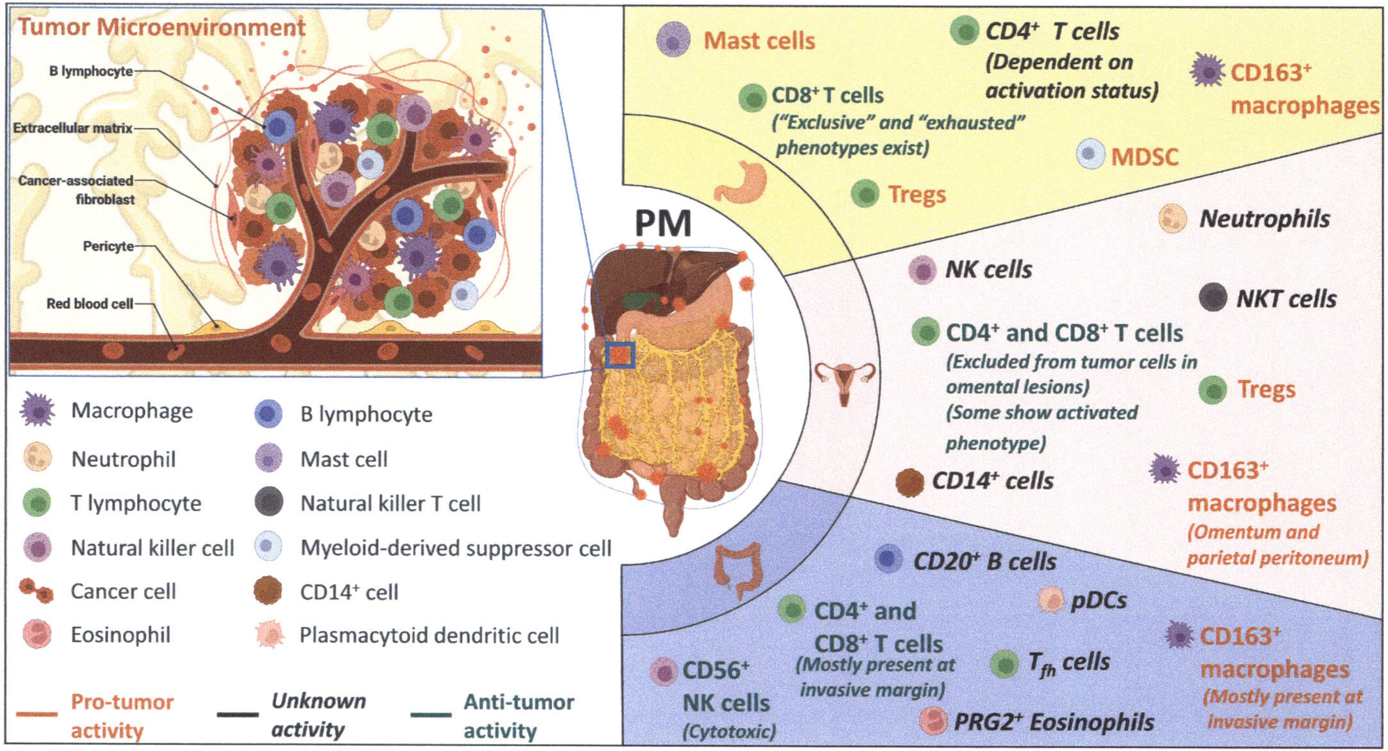

Fig. 3 Overview of the immune contexture of peritoneal metastases in gastric, ovarian, and colorectal cancer, highlighting anti-tumoral (green) vs pro-tumoral (red) effectors.

their receptor CCR1, as well as CD163, a marker for M2-like macrophages. This bulk RNAseq-based pre-metastatic niche formation concept was confirmed by immunofluorescence (IF) and IHC data (Wang et al., 2006). These researchers also studied non-invaded peritoneal samples from malignant and benign OC patients and observed more leukocytes in the former, with the most prominent immune cells being macrophages and T cells. Furthermore, B cells, NK cells and granulocytes were, although marginally present overall, also increased compared to the peritoneum from benign ovarian tumor patients. A significant proportion (range 19–78%) of the macrophages were $CD163^{+}$, thus corresponding to the alternatively activated M2-like macrophages which may thus contribute to the establishment of a pre-metastatic niche. Indeed, others have shown that M2-like macrophages are potent accomplices and may aid tumor cells to adhere and invade the metastatic site by initiating angiogenesis, promoting cellular growth and by remodeling the ECM (Doak et al., 2018; Lin et al., 2019). Some genes responsible for these actions, including PDGFα and β, IGF2R and MMP13, are indeed significantly more expressed in the non-invaded peritoneal samples from malignant OC patients compared to that of benign OC patients (Wang et al., 2005, 2006). Most probably other factors and cell types play a role in priming the pre-metastatic niche, but this was not further explored in these two studies.

Once OC cells detach from the primary tumor and attach to and invade the peritoneum, the interplay between the malignant cells and cells in the ascites and TME prevails. Multiple studies have mapped immune cell subtypes of OC PM using IHC (Dötzer et al., 2019; Ko et al., 2012; Leinster et al., 2012; van Baal et al., 2020) or flow cytometry (Bamias et al., 2007; Tsiatas et al., 2009). An immune cell playing a crucial role in the TME of solid tumors, and most commonly studied in the PM of pOC, is the T cell. Using IHC, Dötzer and colleagues have demonstrated the presence of T cells in omental metastases, and in metastases to the parietal peritoneum (Dötzer et al., 2019). They studied the general $CD3^{+}$ T cell population, as well as the $CD8^{+}$ cytotoxic T cell subpopulation, and observed a significant increase in $CD3^{+}$ and $CD8^{+}$ T cell infiltration in the stroma of the omental metastases compared to the primary tumor. However, no significant differences could be detected between the parietal metastases and the primary tumor, thus emphasizing an anatomical location-driven intrapatient heterogeneity. In addition, the density of $CD3^{+}$, $CD8^{+}$ and PD-1^{+} cells was higher in the stroma compared to the intraepithelial regions of the tumor, possibly indicating an immune excluded phenotype (Chen and Mellman, 2017).

Van Baal et al. correlated this intraepithelial immune infiltration with the intactness of the peritoneal elastic lamina (PEL) in high-grade serous OC PM, and observed a higher density of intraepithelial $CD8^+$ T cells in parietal metastases with an intact PEL (van Baal et al., 2020).

Others have confirmed a prominent T cell infiltration in omental metastases from primary high-grade serous OC using IHC. Leinster and colleagues found that 12% of all omental PM cells stained for $CD3^+$ cells (Leinster et al., 2012). Ko et al., on the other hand, stratified patient samples into $HOXA9^{high}$ and $HOXA9^{low}$ tumors, and showed a reduction in $CD8^+$ T cells and increased Tregs in $HOXA9^{high}$ tumors (Ko et al., 2014). Of note, HOXA9 expression in OC cells was strongly associated with poor survival outcome and correlated with elevated ascites levels of CCL2 and TGF-β2, cytokines which recruit macrophages and induce M2-like TAM polarization, which are potent regulators of Treg and $CD8^+$ T cell recruitment (Ko et al., 2012, 2014). A more in-depth, flow cytometry-based examination of the T cell compartment in PMs highlighted a prominent Treg population of approximately 20% of the $CD4^+$ T cell pool (Bamias et al., 2007; Tsiatas et al., 2009). These authors showed that approximately 40% of $CD4^+$ and 60% of $CD8^+$ cells displayed an activated phenotype, characterized by the expression of HLA-DR and/or CD69. Significantly higher numbers of these activated T cells were observed in serous and endometrioid OC subtypes, compared to the mucinous and clear cell ones, which may in part explain the poorer outcomes observed in patients with the latter subtypes (Lheureux et al., 2019; Tsiatas et al., 2009). Several authors have established a relation between TIL density, therapy response, and survival in OC. In general, platinum-sensitive patients had significantly increased numbers of activated $CD4^+$ T cells in PM compared to platinum-resistant patients (Tsiatas et al., 2009) as well as significantly higher levels of infiltrated $CD8^+$ T cells in omental implants compared to the primary tumor (Dötzer et al., 2019). Furthermore, higher levels of Tregs and activated $CD4^+$ T cells in the TME of PM were associated with improved overall and progression-free survival (Tsiatas et al., 2009), as is an increased $CD8^+$ T cell infiltration (van Baal et al., 2020).

Immune cells other than TILs play have been studied in OC. Based on IHC analysis, Leinster and colleagues found that macrophages accounted for 14% of the total number of cells in omental tumor implants (Leinster et al., 2012). Others proved that a substantial proportion of these macrophages expressed CD163 and were thus identified as alternatively activated M2-like macrophages. More specifically, Ko et al. observed that

$HOXA9^{high}$ omental metastasis samples ($n=16$) not only contained a higher total number of infiltrating macrophages, but also a higher proportion of M2-like macrophages compared to $HOXA9^{low}$ omental implants (75% vs 40%, respectively) (Ko et al., 2014). Interestingly, PM on the parietal peritoneum also comprised a substantial number of CD163-expressing cells, which shows that induction and presence of M2-like macrophages transcends the inter-anatomical location-driven heterogeneity (van Baal et al., 2020). However, the authors did not clarify which cells specifically expressed the CD163 marker.

In addition, Leinster and colleagues also detected a marginal number of neutrophils in omental metastases from high-grade serous OC (approximately 1% of the total number of cells) by IHC staining of neutrophil elastase (Leinster et al., 2012). Attraction of neutrophils to the peritoneal membrane can in part be instigated by MCs producing CXCL6 and CXCL1, common neutrophil chemoattractants, after exposure to malignant ascites (Shahid et al., 2019; Wang et al., 2005). Finally, NK cells, NKT cells and $CD14^{+}$ cells have also been detected in OC PM (Bamias et al., 2007; Tsiatas et al., 2009). The NK ($CD56^{+}$) and NKT cells ($CD56^{+}CD3^{+}$) accounted for a mean of 16.2% and 7.3% of the total number of infiltrating leukocytes, respectively. The same proportions were observed in ascites from these patients (Bamias et al., 2007). Furthermore, $CD56^{+}$ cells were significantly more abundant in the PM of serous and endometrioid OC compared to mucinous and clear cell OC (Tsiatas et al., 2009). Also, CD14, a marker for cells of the mononuclear phagocytic system (i.e. DCs, monocytes and macrophages) was detected on 42% of the infiltrating immune cells in 20 OC PM samples (Bamias et al., 2007).

To sum up, the pre-metastatic peritoneum of OC patients is primed to support the development of peritoneal metastasis. Furthermore, in the resulting omental and parietal PM, a pronounced T cell population is present in the TME, comprising both immunosuppressive and immunostimulatory subpopulations such as Tregs and activated $CD4^{+}$ and $CD8^{+}$ T cells, respectively. However, at least in omental PM, the T cell population seems to be predominantly located in the stroma and thus excluded from the tumor cell population, potentially impeding their anti-tumor activity. Another prominent cell type is the macrophage, a substantial fraction of which displays an alternatively activated phenotype with immunosuppressive properties. Finally, NK cells, NKT cells and a small number of neutrophils are also present in the TME of OC PM (Fig. 3).

5. Ongoing research

Attempts at therapeutic targeting of the TME in phase I trials in solid tumors have not been universally successful, and therefore the implications of interfering within the complex network of interactions taking place in the tumor stroma should be carefully considered (Ramanathan et al., 2019). Recently, a study of cancer vaccination using DCs in the treatment of PM failed to demonstrate a survival benefit, despite this technique already leading to an FDA-approved drug in the treatment of metastatic prostate cancer (Sipuleucel-T (Provenge)). Ramanathan and colleagues utilized αDC1 tumor vaccination combined with a chemokine modulatory regimen in addition to cytoreductive surgery and HIPEC in 46 patients presenting with appendiceal PM, colorectal PM, and malignant peritoneal mesothelioma. Despite favorable cytokine responses, this trial was terminated early, as interim analysis demonstrated futility. Technical challenges in achieving sufficient tumor antigen due to the mucinous nature of these cancers was noted as a key factor in this outcome (Ramanathan et al., 2021).

However, both the high unmet clinical need and the accessibility of the peritoneum allowing for local therapies make PM a highly attractive target for targeting the TIME, as illustrated by the list of ongoing phase I trials addressing the TIME (Table 1). Additionally, results of the ImmunoPeCa trial (NCT00769405) are promising, with a 3-year OS of 72% in colorectal PM, albeit on limited patient numbers. In the ImmunoPeCa trial, a novel immunotoxin MOC31PE, consisting of a linked monoclonal epithelial cell adhesion molecule (EpCAM) antibody (MOC31) with *Pseudomonas* exotoxin A, was administered intraperitoneally following cytoreductive surgery and HIPEC with mitomycin C, with limited additional toxicity (Frøysnes et al., 2021). Further analysis of the patients treated in this trial pointed towards a highly localized immune response, with limited systemic responses. Increased levels of IL-6, IL-1β and TNF all point towards possible immunogenic cell death (Thorgersen et al., 2021).

Several studies utilizing adoptive cell therapy are underway for the treatment of PM. Previous reports of this therapeutic modality in advanced ovarian cancer demonstrated modest effects, potentially due to checkpoint molecules interacting with T cell function (Kverneland et al., 2020; Pedersen et al., 2018). In an ongoing study (NCT04611126) by the same group, addition of a LAG3 antibody is hypothesized to increase efficacy.

Table 1 Ongoing clinical trials targeting the tumor immune microenvironment in peritoneal metastases.

Sponsor	Title	Tumor	Target	Clinicaltrials.gov identifier
Sorrento Therapeutics, Inc.	CAR-T intraperitoneal infusions for CEA-expressing adenocarcinoma peritoneal metastases or malignant ascites (IPC)	CEA-expressing PM	CAR-T cells specific for CEA	NCT03682744
West China Hospital, China	Intraperitoneal infusion of EpCAM CAR-T cell in advanced gastric cancer with peritoneal metastasis (WCH-GC-CART)	EpCAM-expressing Gastric PM	CAR-T cells specific for EpCAM	NCT03563326
Sichuan University, China	Dual-targeting HER2 and PD-L1 CAR-T for cancers with pleural or peritoneal metastasis	HER2 positive solid tumor serosal cavity metastases	HER-2/PD-L1 dual-targeting CAR-T cells	NCT04684459
LintonPharm Co., Ltd.	Comparing the efficacy and safety of intra-peritoneal infusion of catumaxomab and treatment of investigator choice in patients with advanced gastric carcinoma with peritoneal metastasis	Gastric cancer PM	Catumaxomab (a trifunctional antibody targeting EpCAM and CD3)	NCT04222114

Zhejiang Cancer Hospital, China	HIPEC combined with Camrelizumab, paclitaxel and S-1 for conversion therapy in patients with advanced gastric cancer with peritoneal metastasis	Gastric cancer PM	Combination therapy of HIPEC (paclitaxel) and chemotherapy, with added anti PD-1 treatment (Camrelizumab)	NCT04889768
Capital Medical University, China	Hyperthermia combined with immunotherapy in the treatment of abdominal and pelvic malignancies or metastases	Abdominal and pelvic malignancies or metastases	Cytokine-induced Killer (CIK) therapy with/without anti-PD-1 antibody (pembrolizumab)	NCT03757858
Peking University, China	Trial of Surufatinib combined with toripalimab in the treatment of peritoneal metastatic carcinoma of gastrointestinal or primary peritoneal cancer	Gastrointestinal PM and primary peritoneal malignancies	Sulfatinib (targeting VEGFR1-3 and FGFR1) combined with the humanized anti-PD-1 monoclonal antibody Toripalimab	NCT05030246
Herlev Hospital, Denmark	T-cell therapy in combination with Nivolumab, Relatlimab and Ipilimumab for patients with metastatic ovarian cancer	Ovarian cancer PM	Combination therapy of T-cell therapy, combined with anti-PD1 (nivolumab), anti-CTLA-4 (ipilimumab) and anti-LAG3 (relatlimab)	NCT04611126

Both the choice of target of CAR T cells (studies utilizing CEA and EpCAM are underway) and the potentiation of antitumor responses by targeted checkpoint inhibition might prove to be crucial in enabling potent effects.

Further preclinical research points towards other interesting avenues of treatment which have not yet been evaluated in phase I trials. Several groups are targeting the innate immune system or utilizing local therapies targeting the TIME in an attempt to treat PM.

In a syngeneic model of CRC PM, a Korean group demonstrated pronounced M2-like TAM increase and impairment of effector T cell functions. Targeting STING, a main innate immune sensor, was able to turn the "cold" TIME environment into a "hot" TIME environment, leading to effective antitumor response synergistic with PD-1 and/or COX-2 blockade (Lee et al., 2021).

Other local therapies target the adaptive immunity. Si and colleagues reported results of an immunostimulatory antibody (OX40), known to activate T cells, linked to tannic acid in a sprayable gel. Local treatment with this gel resulted in a strong adaptive immune response and was able to induce changes in the TIME of metastatic implants such as decreased MDSC levels and macrophage phenotype switching. These changes resulted in complete inhibition of tumor relapse in a subcutaneous model, and a marked inhibition of growth of a PM model (Si et al., 2021). When utilizing Resiquimod (an FDA-approved TLR7/8 agonist) in a liposomal formulation Pauli and colleagues demonstrated effective inhibition of growth in a murine PM model, and demonstrated synergistic effects when therapy was combined with intraperitoneal oxaliplatin (Pauli et al., 2021). Miller and coworkers showed that inoculation of mice with the virus-like particle CMP-001 (a TLR9 agonist) induces activation of pDCs, coupled with increased infiltration of $CD8^+$ T cells and NK cells and antitumor efficacy in a CRC PM model (Miller et al., 2021).

Additionally, oncolytic virus therapy has shown promising results in an immunocompetent mouse model of CRC PM. Lee and colleagues reported that treatment with JX (an oncolytic vaccinia virus) was able to suppress PM formation. JX was able to selectively infect and kill peritoneal colon cancer cells. Interestingly, JX treatment additionally activated peritoneal DCs and $CD8^+$ T cells. These changes influence the TIME towards antitumor responses (Lee et al., 2020).

6. Conclusion and future perspectives

In the era of immune targeted therapy, elucidation of the immune contexture of peritoneal metastases has received relatively little attention.

Several conclusions may be drawn from the available literature. First, considerable heterogeneity in TIME is evident between cancer (sub)types and, within the same cancer type, between PM and its originating primary cancer. Second, in keeping with the "seed and soil" paradigm, important differences in TIME can exist between several anatomical locations within the same patient (e.g. omentum, parietal peritoneum, visceral peritoneum,...). However, it should be noted that studies based on a single marker to identify immune cell subsets may lead to biased or incomplete results. As an example, some studies used CD56 expression as a marker for NK cells, although it is also expressed by other immune cells such as alpha beta T cells, gamma delta T cells, DCs, and monocytes (van Acker et al., 2017). In addition, the use of the CD4 and CD25 markers alone to annotate Tregs is insufficient, as activated effector T cells also express both markers, and these should ideally be supplemented with the transcription factor FOXP3 and CD127 (Shevyrev and Tereshchenko, 2020). Similarly, CD163 alone is insufficient to distinguish normal macrophages from the alternatively activated M2-like macrophages (Jayasingam et al., 2020). To fully grasp the TIME of PM, a more in depth analysis of PM using detailed -omics of PM is warranted, allowing to identify more cell types and subpopulations such as DCs (cDC1s, cDC2s, pDCs), $CD4^+$ T cells (T_h1, T_h2, T_h17), B cells (B1, B2), granulocytes, and MDSCs as well as their activation and exhaustion statuses.

Clearly, there is a need for detailed morphological and functional analyses of the TIME of PM in clinical samples using advanced platforms such as single cell sequencing and spatial transcriptomics. The insights from these studies may form the biological rationale for novel immune targeting or modulating therapies for patients with PM.

Acknowledgment

Wim Ceelen is a senior clinical researcher from the Fund for Scientific Research – Flanders (FWO). Jesse Demuytere is a PhD fellow fundamental research from the Fund for Scientific Research–Flanders (FWO).

Conflict of interest

The authors do not have any conflict of interest to declare. No funding was received.

References

Albanese, A.M., Albanese, E.F., Miño, J.H., Gómez, E., Gómez, M., Zandomeni, M., et al., 2009. Peritoneal surface area: measurements of 40 structures covered by peritoneum: correlation between total peritoneal surface area and the surface calculated by formulas. Surg. Radiol. Anat. 31, 369–377. https://doi.org/10.1007/S00276-008-0456-9.

Bamias, A., Tsiatas, M.L., Kafantari, E., Liakou, C., Rodolakis, A., Voulgaris, Z., et al., 2007. Significant differences of lymphocytes isolated from ascites of patients with ovarian cancer compared to blood and tumor lymphocytes. Association of CD3+CD56+ cells with platinum resistance. Gynecol. Oncol. 106, 75–81. https://doi.org/10.1016/j.ygyno.2007.02.029.

Boursier, L., Montalto, S.A., Raju, S., Culora, G., Spencer, J., 2006. Characterization of cells of the B lineage in the human adult greater omentum. Immunology 119, 90–97. https://doi.org/10.1111/J.1365-2567.2006.02411.X.

Boutilier, A.J., Elsawa, S.F., 2021. Macrophage polarization states in the tumor microenvironment. Int. J. Mol. Sci. 22, 6995. https://doi.org/10.3390/ijms22136995.

Buechler, M.B., Kim, K.W., Onufer, E.J., Williams, J.W., Little, C.C., Dominguez, C.X., et al., 2019. A stromal niche defined by expression of the transcription factor WT1 mediates programming and homeostasis of cavity-resident macrophages. Immunity 51, 119–130.e5. https://doi.org/10.1016/J.IMMUNI.2019.05.010.

Burg, L., Timmermans, M., van der Aa, M., Boll, D., Rovers, K., de Hingh, I., et al., 2020. Incidence and predictors of peritoneal metastases of gynecological origin: a population-based study in the Netherlands. J. Gynecol. Oncol. 31, 1–12. https://doi.org/10.3802/JGO.2020.31.E58.

Carlow, D.A., Gold, M.R., Ziltener, H.J., 2009. Lymphocytes in the peritoneum home to the omentum and are activated by resident dendritic cells. J. Immunol. 183, 1155–1165. https://doi.org/10.4049/JIMMUNOL.0900409.

Ceelen, W., Ramsay, R.G., Narasimhan, V., Heriot, A.G., de Wever, O., 2020. Targeting the tumor microenvironment in colorectal peritoneal metastases. Trends Cancer 6, 236–246. https://doi.org/10.1016/J.TRECAN.2019.12.008.

Chen, D.S., Mellman, I., 2017. Elements of cancer immunity and the cancer-immune set point. Nature 541, 321–330. https://doi.org/10.1038/NATURE21349.

Demuytere, J., Ceelen, W., van Dorpe, J., Hoorens, A., 2020. The role of the peritoneal microenvironment in the pathogenesis of colorectal peritoneal carcinomatosis. Exp. Mol. Pathol. 115. https://doi.org/10.1016/J.YEXMP.2020.104442.

Doak, G.R., Schwertfeger, K.L., Wood, D.K., 2018. Distant relations: macrophage functions in the metastatic niche. Trends Cancer 4, 445–459. https://doi.org/10.1016/J.TRECAN.2018.03.011.

Dötzer, K., Schlüter, F., Bo Schoenberg, M., Bazhin, A.V., von Koch, F.E., Schnelzer, A., et al., 2019. Immune heterogeneity between primary tumors and corresponding metastatic lesions and response to platinum therapy in primary ovarian cancer. Cancer 11. https://doi.org/10.3390/CANCERS11091250.

Duan, Q., Zhang, H., Zheng, J., Zhang, L., 2020. Turning cold into hot: firing up the tumor microenvironment. Trends Cancer 6, 605–618. https://doi.org/10.1016/j.trecan.2020.02.022.

Frøysnes, I.S., Andersson, Y., Larsen, S.G., Davidson, B., Øien, J.M.T., Julsrud, L., et al., 2021. ImmunoPeCa trial: long-term outcome following intraperitoneal MOC31PE immunotoxin treatment in colorectal peritoneal metastasis. Eur. J. Surg. Oncol. 47, 134–138. https://doi.org/10.1016/J.EJSO.2019.04.014.

Fujimori, D., Kinoshita, J., Yamaguchi, T., Nakamura, Y., Gunjigake, K., Ohama, T., et al., 2020. Established fibrous peritoneal metastasis in an immunocompetent mouse model similar to clinical immune microenvironment of gastric cancer. BMC Cancer 20. https://doi.org/10.1186/S12885-020-07477-X.

Gajewski, T.F., Schreiber, H., Fu, Y.-X., 2013. Innate and adaptive immune cells in the tumor microenvironment. Nat. Immunol. 14, 1014–1022. https://doi.org/10.1038/ni.2703.

Galon, J., Bruni, D., 2019. Approaches to treat immune hot, altered and cold tumours with combination immunotherapies. Nat. Rev. Drug. Discov. 18, 197–218. https://doi.org/10.1038/s41573-018-0007-y.

Galon, J., Pagès, F., Marincola, F.M., Thurin, M., Trinchieri, G., Fox, B.A., et al., 2012. The immune score as a new possible approach for the classification of cancer. J. Transl. Med. 10, 1. https://doi.org/10.1186/1479-5876-10-1.

Gunaydin, G., 2021. CAFs interacting with TAMs in tumor microenvironment to enhance tumorigenesis and immune evasion. Front. Oncol. 11, 668349. https://doi.org/10.3389/fonc.2021.668349. PMID: 34336660; PMCID: PMC8317617.

Gunjigake, K., Kinoshita, J., Yamaguchi, T., Saito, H., Fujimori, D., Horiike, T., et al., 2021. Interleukin-17A derived from mast cells contributes to fibrosis in gastric cancer with peritoneal dissemination. Gastric Cancer 24, 31–44. https://doi.org/10.1007/S10120-020-01092-2.

Haas, K.M., 2015. B-1 lymphocytes in mice and nonhuman primates. Ann. N. Y. Acad. Sci. 1362, 98–109. https://doi.org/10.1111/NYAS.12760.

Hastings, W.D., Tumang, J.R., Behrens, T.W., Rothstein, T.L., 2006. Peritoneal B-2 cells comprise a distinct B-2 cell population with B-1b-like characteristics. Eur. J. Immunol. 36, 1114–1123. https://doi.org/10.1002/EJI.200535142.

Hinshaw, D.C., Shevde, L.A., 2019. The tumor microenvironment innately modulates cancer progression. Cancer Res. 79, 4557–4566. https://doi.org/10.1158/0008-5472.CAN-18-3962.

Jackson-Jones, L.H., Smith, P., Portman, J.R., Magalhaes, M.S., Mylonas, K.J., Vermeren, M.M., et al., 2020. Stromal cells covering omental fat-associated lymphoid clusters trigger formation of neutrophil aggregates to capture peritoneal contaminants. Immunity 52, 700–715.e6. https://doi.org/10.1016/J.IMMUNI.2020.03.011.

Jayasingam, S.D., Citartan, M., Thang, T.H., Mat Zin, A.A., Ang, K.C., Ch'ng, E.S., 2020. Evaluating the polarization of tumor-associated macrophages into M1 and M2 phenotypes in human cancer tissue: technicalities and challenges in routine clinical practice. Front. Oncol. 9, 1512. https://doi.org/10.3389/fonc.2019.01512.

Jiang, Y., Li, Y., Zhu, B., 2015. T-cell exhaustion in the tumor microenvironment. Cell Death Dis. 6, e1792. https://doi.org/10.1038/cddis.2015.162.

Jin, M.-Z., Jin, W.-L., 2020. The updated landscape of tumor microenvironment and drug repurposing. Signal Transduct. Target. Ther. 5, 166. https://doi.org/10.1038/s41392-020-00280-x.

Kim, K.W., Williams, J.W., Wang, Y.T., Ivanov, S., Gilfillan, S., Colonna, M., et al., 2016. MHC II+ resident peritoneal and pleural macrophages rely on IRF4 for development from circulating monocytes. J. Exp. Med. 213, 1951. https://doi.org/10.1084/JEM.20160486.

Ko, S.Y., Barengo, N., Ladanyi, A., Lee, J.-S., Marini, F., Lengyel, E., et al., 2012. HOXA9 promotes ovarian cancer growth by stimulating cancer-associated fibroblasts. J. Clin. Invest. 122, 3603–3617. https://doi.org/10.1172/JCI62229.

Ko, S.Y., Ladanyi, A., Lengyel, E., Naora, H., 2014. Expression of the homeobox gene HOXA9 in ovarian cancer induces peritoneal macrophages to acquire an M2 tumor-promoting phenotype. Am. J. Pathol. 184, 271–281. https://doi.org/10.1016/J.AJPATH.2013.09.017.

Krist, L.F., Koenen, H., Calame, W., van der Harten, J.J., van der Linden, J.C., Eestermans, I.L., et al., 1997. Ontogeny of milky spots in the human greater omentum: an immunochemical study. Anat. Rec. 249. https://doi.org/10.1002/(SICI)1097-0185(199711)249:3<399::AID-AR11>3.0.CO;2-J.

Kverneland, A.H., Pedersen, M., Wulff Westergaard, M.C., Nielsen, M., Borch, T.H., Olsen, L.R., et al., 2020. Adoptive cell therapy in combination with checkpoint inhibitors in ovarian cancer. Oncotarget 11, 2092–2105. https://doi.org/10.18632/ONCOTARGET.27604.

Ladanyi, A., Mukherjee, A., Kenny, H.A., Johnson, A., Mitra, A.K., Sundaresan, S., et al., 2018. Adipocyte-induced CD36 expression drives ovarian cancer progression and metastasis. Oncogene 37, 2285–2301. https://doi.org/10.1038/S41388-017-0093-Z.

Lee, W.J., Ko, S.Y., Mohamed, M.S., Kenny, H.A., Lengyel, E., Naora, H., 2019. Neutrophils facilitate ovarian cancer premetastatic niche formation in the omentum. J. Exp. Med. 216, 176–194. https://doi.org/10.1084/JEM.20181170.

Lee, Y.S., Lee, W.S., Kim, C.W., Lee, S.J., Yang, H., Kong, S.J., et al., 2020. Oncolytic vaccinia virus reinvigorates peritoneal immunity and cooperates with immune checkpoint inhibitor to suppress peritoneal carcinomatosis in colon cancer. J. Immunother. Cancer 8. https://doi.org/10.1136/JITC-2020-000857.

Lee, S.J., Yang, H., Kim, W.R., Lee, Y.S., Lee, W.S., Kong, S.J., et al., 2021. STING activation normalizes the intraperitoneal vascular-immune microenvironment and suppresses peritoneal carcinomatosis of colon cancer. J. Immunother. Cancer 9. https://doi.org/10.1136/JITC-2020-002195.

Lei, X., Lei, Y., Li, J.-K., Du, W.-X., Li, R.-G., Yang, J., et al., 2020. Immune cells within the tumor microenvironment: biological functions and roles in cancer immunotherapy. Cancer Lett. 470, 126–133. https://doi.org/10.1016/j.canlet.2019.11.009.

Leinster, D.A., Kulbe, H., Everitt, G., Thompson, R., Perretti, M., Gavins, F.N.E., et al., 2012. The peritoneal tumour microenvironment of high-grade serous ovarian cancer. J. Pathol. 227, 136–145. https://doi.org/10.1002/path.4002.

Lemoine, L., Sugarbaker, P., van der Speeten, K., 2016. Pathophysiology of colorectal peritoneal carcinomatosis: Role of the peritoneum. World J. Gastroenterol. 22 (34), 7692–7707. https://doi.org/10.3748/wjg.v22.i34.7692.

Lheureux, S., Braunstein, M., Oza, A.M., 2019. Epithelial ovarian cancer: evolution of management in the era of precision medicine. CA Cancer J. Clin. 69. https://doi.org/10.3322/CAAC.21559.

Li, M.O., Wolf, N., Raulet, D.H., Akkari, L., Pittet, M.J., Rodriguez, P.C., et al., 2021. Innate immune cells in the tumor microenvironment. Cancer Cell 39, 725–729. https://doi.org/10.1016/j.ccell.2021.05.016.

Liao, C.T., Andrews, R., Wallace, L.E., Khan, M.W.A., Kift-Morgan, A., Topley, N., et al., 2017. Peritoneal macrophage heterogeneity is associated with different peritoneal dialysis outcomes. Kidney Int. 91, 1088–1103. https://doi.org/10.1016/J.KINT.2016.10.030.

Lin, Y., Xu, J., Lan, H., 2019. Tumor-associated macrophages in tumor metastasis: biological roles and clinical therapeutic applications. J. Hematol. Oncol. 12. https://doi.org/10.1186/S13045-019-0760-3.

Liu, J., Geng, X., Li, Y., 2016. Milky spots: omental functional units and hotbeds for peritoneal cancer metastasis. Tumour Biol. 37, 5715–5726. https://doi.org/10.1007/S13277-016-4887-3.

Michailova, K.N., Wassilev, W.A., Kühnel, W., 2005. Features of the peritoneal covering of the lesser pelvis with special reference to stomata regions. Ann. Anat. 187, 23–33. https://doi.org/10.1016/J.AANAT.2004.05.002.

Mikuła-Pietrasik, J., Uruski, P., Tykarski, A., Krzysztof, K., 2018. The peritoneal "soil" for a cancerous "seed": a comprehensive review of the pathogenesis of intraperitoneal cancer metastases. Cell. Mol. Life Sci. 75, 509–525. https://doi.org/10.1007/s00018-017-2663-1.

Miller, A.M., Lemke-Miltner, C.D., Blackwell, S., Tomanek-Chalkley, A., Gibson-Corely, K.N., Coleman, K.L., et al., 2021. Intraperitoneal CMP-001: a novel immunotherapy for treating peritoneal carcinomatosis of gastrointestinal and pancreaticobiliary cancer. Ann. Surg. Oncol. 28, 1187–1197. https://doi.org/10.1245/S10434-020-08591-7.

Mutsaers, S.E., Prêle, C.M.A., Pengelly, S., Herrick, S.E., 2016. Mesothelial cells and peritoneal homeostasis. Fertil. Steril. 106, 1018–1024. https://doi.org/10.1016/J.FERTNSTERT.2016.09.005.

Nisitani, S., Murakami, M., Akamizu, T., Okino, T., Ohmori, K., Mori, T., et al., 1997. Preferential localization of human CD5+ B cells in the peritoneal cavity. Scand. J. Immunol. 46, 541–545. https://doi.org/10.1046/J.1365-3083.1997.D01-166.X.

Ohtani, O., Ohtani, Y., Li, R.X., 2001. Phylogeny and ontogeny of the lymphatic stomata connecting the pleural and peritoneal cavities with the lymphatic system—a review. Ital. J. Anat. Embryol. 106.

Pardoll, D.M., 2012. The blockade of immune checkpoints in cancer immunotherapy. Nat. Rev. Cancer 12, 252–264. https://doi.org/10.1038/nrc3239.

Pauli, G., Chao, P.-H., Qin, Z., Böttger, R., Lee, S.E., Li, S.-D., 2021. Liposomal resiquimod for enhanced immunotherapy of peritoneal metastases of colorectal cancer. Pharmaceutics 13, 1696. https://doi.org/10.3390/PHARMACEUTICS13101696.

Pedersen, M., Westergaard, M.C.W., Milne, K., Nielsen, M., Borch, T.H., Poulsen, L.G., et al., 2018. Adoptive cell therapy with tumor-infiltrating lymphocytes in patients with metastatic ovarian cancer: a pilot study. Oncoimmunology 7. https://doi.org/10.1080/2162402X.2018.1502905.

Perez-Shibayama, C., Gil-Cruz, C., Cheng, H.W., Onder, L., Printz, A., Mörbe, U., et al., 2018. Fibroblastic reticular cells initiate immune responses in visceral adipose tissues and secure peritoneal immunity. Sci. Immunol. 3. https://doi.org/10.1126/SCIIMMUNOL.AAR4539.

Pesce, S., Belgrano, V., Greppi, M., Carlomagno, S., Squillario, M., Barla, A., et al., 2019. Different features of tumor-associated NK cells in patients with low-grade or high-grade peritoneal carcinomatosis. Front. Immunol. 10. https://doi.org/10.3389/FIMMU.2019.01963.

Ramanathan, R.K., McDonough, S.L., Philip, P.A., Hingorani, S.R., Lacy, J., Kortmansky, J.S., et al., 2019. Phase IB/II randomized study of FOLFIRINOX plus pegylated recombinant human hyaluronidase versus FOLFIRINOX alone in patients with metastatic pancreatic adenocarcinoma: SWOG S1313. J. Clin. Oncol. 37, 1062–1069. https://doi.org/10.1200/JCO.18.01295.

Ramanathan, R., Choudry, H., Jones, H., Girgis, M., Gooding, W., Kalinski, P., et al., 2021. Phase II trial of adjuvant dendritic cell vaccine in combination with celecoxib, interferon-α, and rintatolimod in patients undergoing cytoreductive surgery and hyperthermic intraperitoneal chemotherapy for peritoneal metastases. Ann. Surg. Oncol. 28, 4637–4646. https://doi.org/10.1245/S10434-020-09464-9.

Rosas, M., Davies, L.C., Giles, P.J., Liao, C.T., Kharfan, B., Stone, T.C., et al., 2014. The transcription factor Gata6 links tissue macrophage phenotype and proliferative renewal. Science (New York, NY) 344, 645–648. https://doi.org/10.1126/SCIENCE.1251414.

Rynne-Vidal, A., Jiménez-Heffernan, J.A., Fernández-Chacón, C., López-Cabrera, M., Sandoval, P., 2015. The mesothelial origin of carcinoma associated-fibroblasts in peritoneal metastasis. Cancer 7, 1994–2011. https://doi.org/10.3390/CANCERS7040872.

Seebauer, C.T., Brunner, S., Glockzin, G., Piso, P., Ruemmele, P., Schlitt, H.J., et al., 2016. Peritoneal carcinomatosis of colorectal cancer is characterized by structural and functional reorganization of the tumor microenvironment inducing senescence and proliferation arrest in cancer cells. OncoImmunology 5. https://doi.org/10.1080/2162402X.2016.1242543/SUPPL_FILE/KONI_A_1242543_SM3568.ZIP.

Shahid, S., Iman, A., Matti, U., Rachid, K., Assaf, A., Eveno, C., et al., 2019. Fibrin deposit on the peritoneal surface serves as a niche for cancer expansion in carcinomatosis patients. Neoplasia (New York, NY) 21, 1091–1101. https://doi.org/10.1016/J.NEO.2019.08.006.

Shevyrev, D., Tereshchenko, V., 2020. Treg heterogeneity, function, and homeostasis. Front. Immunol. 10, 3100. https://doi.org/10.3389/fimmu.2019.03100.

Shimotsuma, M., Takahashi, T., Kawata, M., Dux, K., 1991. Cellular subsets of the milky spots in the human greater omentum. Cell Tissue Res. 264, 599–601. https://doi.org/10.1007/BF00319049.

Si, X., Ji, G., Ma, S., Xu, Y., Zhao, J., Zhang, Y., et al., 2021. In-situ-sprayed dual-functional immunotherapeutic gel for colorectal cancer postsurgical treatment. Adv. Healthc. Mater. 10. https://doi.org/10.1002/ADHM.202100862.

Sommerfeld, L., Finkernagel, F., Jansen, J.M., Wagner, U., Nist, A., Stiewe, T., et al., 2021. The multicellular signalling network of ovarian cancer metastases. Clin. Transl. Med. 11. https://doi.org/10.1002/CTM2.633.

Song, H., Wang, T., Tian, L., Bai, S., Chen, L., Zuo, Y., et al., 2019. Macrophages on the peritoneum are involved in gastric cancer peritoneal metastasis. J. Cancer 10, 5377–5387. https://doi.org/10.7150/JCA.31787.

Sugarbaker, P.H., 2018. Gastric cancer: prevention and treatment of peritoneal metastases. J. Cancer Metastasis Treatment 4, 7. https://doi.org/10.20517/2394-4722.2017.67.

Thorgersen, E.B., Asvall, J., Frøysnes, I.S., Schjalm, C., Larsen, S.G., Dueland, S., et al., 2021. Increased local inflammatory response to MOC31PE immunotoxin after cytoreductive surgery and hyperthermic intraperitoneal chemotherapy. Ann. Surg. Oncol. 28, 5252–5262. https://doi.org/10.1245/S10434-021-10022-0.

Torre, L.A., Trabert, B., DeSantis, C.E., Miller, K.D., Samimi, G., Runowicz, C.D., et al., 2018. Ovarian cancer statistics, 2018. CA Cancer J. Clin. 68, 284–296. https://doi.org/10.3322/caac.21456.

Trim, W.V., Lynch, L., 2021. Immune and non-immune functions of adipose tissue leukocytes. Nat. Rev. Immunol. https://doi.org/10.1038/S41577-021-00635-7.

Tsiatas, M.L., Gyftaki, R., Liacos, C., Politi, E., Rodolakis, A., Dimopoulos, M.A., et al., 2009. Study of T lymphocytes infiltrating peritoneal metastases in advanced ovarian cancer: associations with vascular endothelial growth factor levels and prognosis in patients receiving platinum-based chemotherapy. Int. J. Gynecol. Cancer 19, 1329–1334. https://doi.org/10.1111/IGC.0B013E3181B7A40E.

van Acker, H.H., Capsomidis, A., Smits, E.L., van Tendeloo, V.F., 2017. CD56 in the immune system: more than a marker for cytotoxicity? Front. Immunol. 8, 892. https://doi.org/10.3389/fimmu.2017.00892.

van Baal, J.O.A.M., Lok, C.A.R., Jordanova, E.S., Horlings, H., van Driel, W.J., Amant, F.C., et al., 2020. The effect of the peritoneal tumor microenvironment on invasion of peritoneal metastases of high-grade serous ovarian cancer and the impact of neoadjuvant chemotherapy. Virchows Archiv. Int. J. Pathol. 477, 535–544. https://doi.org/10.1007/S00428-020-02795-8.

van Driel, W.J., Koole, S.N., Sikorska, K., van Leeuwen, J.H., Schreuder, H.W.R., Hermans, R.H.M., et al., 2018. Hyperthermic intraperitoneal chemotherapy in ovarian cancer. N. Engl. J. Med. 378, 230–240. https://doi.org/10.1056/NEJMoa1708618.

Wang, E., Ngalame, Y., Panelli, M.C., Nguyen-Jackson, H., Deavers, M., Mueller, P., et al., 2005. Peritoneal and subperitoneal stroma may facilitate regional spread of ovarian cancer. Clin. Cancer Res. 11, 113–122.

Wang, X., Deavers, M., Patenia, R., Bassett, R.L.J., Mueller, P., Ma, Q., et al., 2006. Monocyte/macrophage and T-cell infiltrates in peritoneum of patients with ovarian cancer or benign pelvic disease. J. Transl. Med. 4, 30. https://doi.org/10.1186/1479-5876-4-30.

Wang, R., Song, S., Harada, K., Ghazanfari Amlashi, F., Badgwell, B., Pizzi, M.P., et al., 2020. Multiplex profiling of peritoneal metastases from gastric adenocarcinoma identified novel targets and molecular subtypes that predict treatment response. Gut 69, 18–31. https://doi.org/10.1136/GUTJNL-2018-318070.

Wei, C., Yang, C., Wang, S., Shi, D., Zhang, C., Lin, X., et al., 2019. Crosstalk between cancer cells and tumor associated macrophages is required for mesenchymal circulating tumor cell-mediated colorectal cancer metastasis. Mol. Cancer 18, 64. https://doi.org/10.1186/s12943-019-0976-4.

Wu, D., Han, J.M., Yu, X., Lam, A.J., Hoeppli, R.E., Pesenacker, A.M., et al., 2019. Characterization of regulatory T cells in obese omental adipose tissue in humans. Eur. J. Immunol. 49, 336–347. https://doi.org/10.1002/EJI.201847570.

Yamaguchi, T., Fushida, S., Yamamoto, Y., Tsukada, T., Kinoshita, J., Oyama, K., et al., 2016. Tumor-associated macrophages of the M2 phenotype contribute to progression in gastric cancer with peritoneal dissemination. Gastric Cancer 19, 1052–1065. https://doi.org/10.1007/S10120-015-0579-8/FIGURES/5.

Yamaguchi, T., Fushida, S., Kinoshita, J., Okazaki, M., Ishikawa, S., Ohbatake, Y., et al., 2020. Extravasated platelet aggregation contributes to tumor progression via the accumulation of myeloid-derived suppressor cells in gastric cancer with peritoneal metastasis. Oncol. Lett. 20, 1879–1887. https://doi.org/10.3892/OL.2020.11722.

Yamaguchi, T., Kinoshita, J., Saito, H., Shimada, M., Terai, S., Moriyama, H., et al., 2021. High CD8/CD33 ratio in peritoneal metastatic lesions is associated with favorable prognosis in gastric cancer. Cancer Rep. (Hoboken, NJ) 4. https://doi.org/10.1002/CNR2.1389.

Yonemura, Y., Ishibashi, H., Mizumoto, A., Nishihara, K., Liu, Y., Wakama, S., et al., 2020. Mechanisms of peritoneal metastasis formation. Pathol. Perit. Metastases, 1–26. https://doi.org/10.1007/978-981-15-3773-8_1.

Yung, S., Li, F.K., Chan, T.M., 2006. Peritoneal mesothelial cell culture and biology. Perit. Dial. Int. 26, 162–173. https://doi.org/10.1177/089686080602600207.

Zhang, N., Kim, S.H., Gainullina, A., Erlich, E.C., Onufer, E.J., Kim, J., et al., 2021. LYVE1+ macrophages of murine peritoneal mesothelium promote omentum-independent ovarian tumor growth. J. Exp. Med. 218. https://doi.org/10.1084/JEM.20210924.

Zhao, Y., Chen, C., Xu, X., Ge, X., Ding, K., Zheng, S., et al., 2021. An efficient prognostic immune scoring system for colorectal cancer patients with peritoneal metastasis. Oncoimmunology 10. https://doi.org/10.1080/2162402X.2021.1901464.

Zhu, L., Narloch, J.L., Onkar, S., Joy, M., Broadwater, G., Luedke, C., et al., 2019. Metastatic breast cancers have reduced immune cell recruitment but harbor increased macrophages relative to their matched primary tumors. J. Immunother. Cancer 7, 265. https://doi.org/10.1186/s40425-019-0755-1.

CHAPTER FIVE

Immune response and locoregional treatments for peritoneal carcinomatosis

Caterina Fiorentini[a], Donatella Sarti[b], Stefano Guadagni[c], and Giammaria Fiorentini[d,*]

[a]Department of Prevention and Sport Medicine, University Hospital Klinikum rechts der Isar, Technical University of Munich, Munich, Germany
[b]Oncology Department, S. Maria Della Misericordia Hospital, ASUR1, Urbino, Italy
[c]Department of Applied Clinical Sciences and Biotechnology, University of L'Aquila, L'Aquila, Italy
[d]Department of Onco-Hematology, Azienda Ospedaliera "Ospedali Riuniti Marche Nord", Pesaro, Italy
*Corresponding author: e-mail address: g.fiorentini@alice.it

Contents

Abstract

Peritoneal Carcinomatosis (PC) is considered as a terminal disease with short survival. It is treated with palliative therapies, consisting of repeated drainages and sometimes instillation of chemotherapy. Since the nineties, surgery has been combined with more effective systemic chemotherapy, intraperitoneal chemotherapy and hyperthermic intraperitoneal chemotherapy (HIPEC) for the treatment of PC. This combination

International Review of Cell and Molecular Biology, Volume 371
ISSN 1937-6448
https://doi.org/10.1016/bs.ircmb.2022.04.016

therapy significantly increases the overall survival of selected PC patients. The understanding of how intraperitoneal chemotherapy and HIPEC can cure patients is still unclear. Experts hypothesized that the efficacy is obtained by the ability of high peritoneal drug exposure and hyperthermia to directly kill cancer cells. Several studies indicate that cancer cells death directly influences the response of the immune system. For this reason, the protective effect of intraperitoneal chemotherapy and HIPEC could be mediated by its ability to kill cancer cells in an immuno-genic way, causing an efficient anticancer immune response. In this review, we investigate the role of the innate peritoneal or locoregional therapy-induced immune response in PC therapy.

Abbreviations

ADCC antibody-dependent cellular cytotoxicity
CRS cytoreductive surgeryPCperitoneal carcinomatosis
DAMPs damage associated molecular patterns
EGFR epidermal growth factor
EpCAM human epithelial cell adhesion molecule
HIPEC hyperthermic intra peritoneal chemotherapy
ICD immunogenic cell death
IGFR insulin growth factor receptor
LAK lymphokine activated killer cells
mAbs antibodies
OS overall survival
PF peritoneal fluid
PIPAC pressurized intraperitoneal aerosol chemotherapy
TIL tumor infiltrating lymphocytes
VEGF vascular endothelial growth factor

1. Introduction

Peritoneal carcinomatosis (PC) is represented by late-stage metastases of intra-abdominal malignancies to the peritoneal surface, resulting in dissemination of malignancy in the peritoneal cavity (Yap et al., 2021). PC is associated with a poor prognosis for patients with short survival (3.1 months) and frequent recurrence (4–50%) (Coccolini et al., 2013; Sadeghi et al., 2000). However, the introduction of cytoreductive surgery (CRS) and hyperthermic intraperitoneal chemotherapy (HIPEC) significantly improves survival (up to 50 months of patients with PC from colorectal cancer) and reduction of disease burden in PC patients (Breuer et al., 2021). These improvements of therapeutic outcomes are also linked to earlier diagnosis, improved biological and genetic knowledge, use of the peritoneal cancer index, a standardized scoring system to quantify peritoneal tumor burden during an operation, and the cytoreduction score (CC score) to determine radicality in surgery (Yap et al., 2021).

HIPEC is a localized chemotherapeutic treatment that has several advantages: deep tissue penetration and anti-tumor cell response due to hyperthermia, eliminating residual microscopic tumors post CRS in an adjuvant setting. However, due to the non-negligible side effects of CRS + HIPEC, several authors have reconsidered this approach and investigated the best prognostic factors for patients selection and different methods of HIPEC application (Hallam et al., 2019; Quénet et al., 2021).

Immunotherapy can offer new possibilities of cure in many cancers and its efficacy is not dependent of the cell cycle or chemo-resistance, attacking metastatic cells also in their dormant cell phase, when chemotherapy is not effective. For this reason, several studies aim to study the immune response of patients with PC (Mulder et al., 2019; Mulens-Arias et al., 2021; Thorgersen et al., 2021).

In this review, we investigate the role of the innate peritoneal or locoregional therapy-induced immune response in PC therapy.

1.1 Anatomy and physiology of the peritoneum

The peritoneum is a large serous membrane of mesodermal origin with mesenchymal and epithelial characteristics. It can be divided in parietal and visceral peritoneum and cover respectively the abdomen and the visceral organs. The peritoneum consists of several layers: glycocalix, mesothelium, basal lamina (collagen IV and laminin), submesothelial connective tissue and elastic lamina. The mesodermal layer acts as defense due to the tight junctions among the cells and expresses cytokeratin, and fibronectin (van Baal et al., 2018). The space between the bilaminar layers contains the peritoneal fluid (PF) that facilitates frictionless movement of abdominal organs during peristalsis, removes pathogens and cells and allows reparative events. The PF is highly fibrinolytic and its volume is around 100 mL (Kastelein et al., 2019). The main role of peritoneum is the regulation of intraperitoneal homeostasis of the abdominal cavity, moreover it plays an important role in antigen presentation, inflammatory responses, fibrosis and fibrinolysis, tissue repair and peritoneal metastases development (van Baal et al., 2018).

1.2 Development of peritoneal carcinomatosis

PC is characterized by superficial tumor deposits on the inner surface of the peritoneum; deep invasion is rare. PC originate from the mesothelium or disseminate from other primary tumor sites. The most common type of primary peritoneal cancer is serous carcinoma and accounts for 10% of pelvic tumors. Malignant mesothelioma and pseudomixoma peritoneii are less common but severe tumors. Peritoneal metastases or secondary PC arise from gastrointestinal, gynecological and extra-abdominal tumors (Anwar and Kasi, 2022). PC is present in 75% of ovarian cancers, 5–10% of colorectal tumors, 14% of gastric cancers and 9% of other malignancies (mainly breast and lung) at diagnosis and has metachronous malignant proliferation in 20–50% of cases (Beniey, 2019; Chen et al., 2021; Desiderio et al., 2017; Lemoine et al., 2016; Sereno et al., 2013; van Baal et al., 2018).

The diffusion from primary tumor is due to the intramural growth beyond the serosal layers or spillover after surgical approaches. Cancer cells initial detachment is due to the down-regulation of E- Cadherin (intracellular adhesion molecule) (Lemoine et al., 2016). Tumors cells diffusion to distant sites occurs via vascular, lymphatic, or trans-coelomic route. Vascular factors of peritoneal fluid promote metastatic cancer cell growth during their travel. During lymphatic transport, cancer cells enter the lymph capillaries at sites of immune cell aggregates (milky spots) or lymphatic stomata (Liu et al., 2016).

Metastases origin involve three steps: adhesion, angiogenesis, and immune evasion. Adhesion is mediated by the interaction of tumor cell receptors (CD44) with molecules on mesothelial cells, such as intercellular adhesion molecule 1 (ICAM-1) and vascular adhesion molecule 1 (VCAM-1). This interaction triggers the release of cytokines such as tumor necrosis alpha (TNFα), interleukin- 1beta (IL-1β), and interleukin-1 gamma (IL-1γ) (Lemoine et al., 2016).

After adhesion, cancer cells invade the sub-mesothelial stroma, where metastases growth is mediated by the production of growth factors (IGF-1 and EGFR) and new blood vessels are formed through the production of VEGF and hypoxia-inducible factor (HIF-1) (Capobianco et al., 2017).

1.3 Pharmacological aspects of intraperitoneal drug administration

High intraperitoneal drug concentrations can be obtained when the anticancer agent is administered directly into the peritoneal cavity. This allows to maintain low systemic drug concentrations, due to the "peritoneal–plasma barrier" that slow drug absorption from the peritoneal cavity into systemic blood circulation, resulting in a high cytotoxic drug concentration ratio between peritoneal cavity and plasma. This advantage is represented by area under the concentration (AUC)–time curve gradient. The intraperitoneal/plasma drug AUC ratio varies (10–1,000) according to drugs molecular weight and their hepatic and renal clearance (de Bree et al., 2017). This high intraperitoneal drug concentration and tumor cell exposure constitute are effective in PC eradication. A major downside of intraperitoneal chemotherapy, however, is that drugs cannot penetrate inside the depth of the tumor deposits. In order to address this issue, hyperthermic solution for peritoneal lavage are associated to antineoplastic drugs for PC treatment (Van der Speeten et al., 2009; van Ruth et al., 2003).

1.4 HIPEC and PIPAC: Antineoplastic drugs used

Hyperthermia (>39–40 °C) has a selective cytotoxic effect on cancer cells and hence can increase antineoplastic drugs efficacy. This effect is particularly beneficial in the treatment of PC and is used in intraoperative HIPEC to enhance drug penetration inside tumor nodules, maintaining its pharmacokinetic profile (Van der Speeten et al., 2009; van Ruth et al., 2003). The highest thermal enhancement is observed for melphalan, ifosfamide, platinum compounds, mitomycin C and doxorubicin (Kusamura et al., 2008; van Ruth et al., 2003). Some of these agents, especially anthracyclines, can induce immunogenic cell death, leading to the release of damage-associated molecular patterns (DAMPs) molecules from dying cancer cells. These molecules are a component of the innate immune response and can activate dendritic cells maturation and antigen presentation to CD8+ T cells after antigen release. Antigen-activated T cells clonally expand and can attack surviving cancer cells.

Tesniere and colleagues show that oxaliplatin efficiently triggers high-mobility group box 1 protein release and stimulates pre-apoptotic calreticulin exposure in a series of murine and human colon cancer cell lines that is reduced by short interfering RNA-mediated depletion of calreticulin. They suggest that oxaliplatin is important to induce immunogenic death of tumor cells (Tesniere et al., 2010).

Mitomycin C (10–50 mg/m^2 over 60–120 min at 41–44 °C) and oxaliplatin (460 mg/m^2 over 30 min at 42–44 °C) are the two most used cytotoxic drugs in HIPEC and are usually preceded by an intravenous infusion of 4-FU (400 mg/m^2) with leucovorin (20 mg/m^2). Mitomycin C is more effective than oxaliplatin in colorectal PC treatment with HIPEC, while maintaining similar rates of intra-abdominal complications (Prada-Villaverde et al., 2014). Leung and colleagues, however, show that patients treated with oxaliplatin had better prognoses than those with mitomycin C-based HIPEC with a median survival of 54 vs 26 months (Leung et al., 2017).

A complete response of PC to any treatment is considered a rare occurrence and many studies highlight the palliative aspect of the care. For this reason, new methods have been developed with fewer and side effects and better quality of life. Pressurized Intraperitoneal Aerosol Chemotherapy (PIPAC) is a method that applies chemotherapy intraperitoneally during laparoscopy as an under-pressure airflow of 12 mmHg, usually every four weeks (De Simone et al., 2020). This method is currently used in patients with non-resectable disease and is well tolerated (Solass et al., 2009). The rationale for PIPAC is a more homogeneous intraperitoneal distribution of chemotherapy and the achievement of higher local drug concentrations than HIPEC (Nadiradze et al., 2019; Solass et al., 2009). PIPAC treatment can be repeated, increasing its clinical efficacy and allowing to perform biopsies for molecular and immunological investigations

2. Immunological features of the peritoneum

The peritoneum can induce immunological responses, promote recruitment and proliferation of T lymphocyte induced or mediated specific immunity against bacterial or viral pathogens and cancer cells. In the peritoneal cavity there are different immune competent cells: 42% monocytes and macrophages (CD68+), 40% T-lymphocytes (CD2+), 10% NK-cells and 5 % Dendritic cells (Kastelein et al., 2019). A high percentage of peritoneal CD4+ (92%) and CD8+ (73%) T-lymphocytes are CD45RO+, indicating that they already have differentiated memory and effectors T cell phenotype. Monocytes and macrophages remove of debris and bacteria, while B1 lymphocytes produce natural antibodies (IgM and IgA) with broad specificity and low antigen affinity (Liu et al., 2016).

2.1 Immunotherapy for tumors: Old approaches

The efficacy of immunotherapy is fascinating because it is not related to chemo-resistance or cell cycle and is dependent on antigen recognition and activation of immune effectors cells. For this reason, immunotherapy can efficiently attack cancer cells present of peritoneal cavity also when they are dormant and hence anti-cancer drugs are not effective. During eighty and ninety, drugs with non-specific immunostimulation abilities, such as the streptococcal OK-432, interleukines and antibodies, have been studied.

OK-432 induce stimulatory activity on natural killer cells (NK), lymphokine activated killer cells (LAK) and T-lymphocytes in a rat model (Ono et al., 2006). OK-432 has clinical effect in gastric cancer-derived PC patients, inducing of a predominant Th1 response. This induction and stimulation of innate components (NK-cells and macrophages) can obtain tumor response, increasing median overall survival (42.6 months vs 32.3 months for the control group) with overall hazard ratio of 0.88 (95% confidence interval 0.77–1.00, $P = 0.050$). However, immunochemotherapy treatment with OK-432 may have only a borderline significant effect for patients with stage III or stage IV gastric cancer after curative resection (Oba et al., 2016).

Interleukin-2 (IL-2) is most effective cytokine in immunotherapy (Rosenberg et al., 1989). It stimulates proliferation and killing capacity of T-lymphocytes and activates lymphokine activated killer cells (LAK). Application of LAK cells resulted in significant results in animal models.

The concept of tumor infiltrating lymphocytes (TIL) is introduced in order to differentiate the specific killing mechanisms of differentiated T lymphocytes after contact to cancer cells by effective immune stimulation with IL-2, from the unspecific immune activation. TILs mediate significant killing effects, but are difficult to be expanded in large amount for immunotherapy, moreover, IL-2 regimens induce severe side effects, for this reason they are not used anymore in clinical practice (Rosenberg et al., 1993).

Monoclonal antibodies (mAbs) bind to cancer cell cellular membrane, resulting in immune response activation, tumor cell receptor signaling block, growth inhibition due to cell cycle arrest, apoptosis and anti-angiogenetic effects. mAbs induce also secondary immune effects by recruiting accessory cells via their intact Fc-region, which contribute to immune-response in terms of antibody-dependent cell cytotoxicity (ADCC). The essential problem of the monoclonal antibody is the lack of a direct anti-cancer effectors component, moreover they can only bind to antigen-processing cells like macrophages, but not T-helper cells or cytotoxic T-lymphocytes (Jäger et al., 2001).

2.2 Human epithelial cell adhesion molecule and triple functional antibody

Since 2000s a new approach has been studied in order to identifying neoplastic cells and selectively kill them. The differential recognition of peritoneal cells from tumor cells is possible, the former have mesenchymal origin and do not express antigens related to epithelial tumor cells. For this reason, antibody constructs with a binding site against a specific epithelial antigen can be applied in PC therapy.

Epithelial cell adhesion molecule (EpCAM) is a type of transmembrane glycoprotein that mediates epithelium-specific cell-to-cell adhesion and is identified as a tumor-associated antigen due to its high expression level in rapidly growing epithelial tumors, such as colon, gastric, lung and prostate cancer (Heiss et al., 2005; Went et al., 2006).

Based on this evidence the trifunctional antibodies have been developed. These antibodies have interesting triple feature: binding to specific cancer antigens, binding and activating CD3+ T lymphocytes and recruiting accessory immune cells by their intact Fc-region. Catumaxomab is the first tri-functional monoclonal antibody, consisting of a mouse immunoglobulin G (IgG)2a chain (binding to the human EpCAM), a rat IgG2b chain (binding to human CD3 on T-lymphocytes) and a third functional binding site within the hybrid Fc-region selectively binds to and activates Fcγ-receptor-type I (CD64), -type IIa (CD16) and -type III (CD16) positive accessory cells (Ruf and Lindhofer, 2001).

The trifunctional antibodies presented an efficient suppression of cancer cells by several killing mechanisms: T-cell mediated lysis through perforin and granzyme B, cytotoxicity by T-cell secreted cytokines (IFN-γ and TNF-α) and phagocytosis via activation of accessory cells (Heiss et al., 2005, 2010). Intraperitoneal therapy with trifunctional antibodies increases specific tumor-reactive CD4+ and CD8+ T-lymphocytes in PC patients, destroying cancer cells and inducing specific T-cells mediated immunity (Heiss et al., 2005, 2010).

3. Ongoing research

The peritoneum is considered a pharmacological model for the administration of conventional antiblastic drugs for the main tumors with PC, such as ovarian, stomach and colon cancers. Patients, selection criteria and the advantages and limitations due to toxicity of intraperitoneal chemotherapy are widely known. HIPEC should have been an improvement of classic intraperitoneal drug administration but its utility within subsets of PC is not fully understood yet.

A Cochrane investigation on intraperitoneal chemotherapy for primary epithelial ovarian cancer show that it increases overall survival and progression-free survival. These results are the most reliable estimates of the relative survival benefits of intraperitoneal vs systemic chemotherapy and should be used as part of the decision making process in ovarian cancer therapy. However, the potential for side effects and catheter related complications needs to be considered and optimal dose, timing and mechanism of administration cannot be addressed from this meta-analysis (Jaaback et al., 2016).

As concerning HIPEC, it has a survival advantage in gastric cancer patients and has the most important efficacy in patients with disease burden that is limited to positive cytology and limited nodal involvement (Desiderio et al., 2017). The study PRODIGE 7 analyzed 265 colon cancer patients randomly assigned, 133 to the cytoreductive surgery plus HIPEC and 132 to the cytoreductive surgery alone group. After median follow-up of 63.8 months, HIPEC has similar effect on median overall survival (41.7 months 95% CI 36.2–53.8, vs 41.2 months 95% CI 35.1–49.7) and early stage tolerance, however HIPEC induces more late complications that control group. For this reason, cytoreductive surgery alone should be the cornerstone of therapeutic strategies with curative intent for colorectal peritoneal metastases. In any case, some flaws are reported: first of all the duration of the HIPEC with short exposure time of chemotherapy 30 min, whereas most HIPEC treatments are for 90 min up to 2 h in order to maximally utilize the drug administered directly into the peritoneal space. Moreover with only 30 min the action of hyperthermia would be minimal and perhaps absent.

3.1 Intraperitoneal immunotherapy

On the basis of these observations and the unsatisfactory efficacy of intraperitoneal therapies to improve the prognosis of PC, the attention has shifted to the immunological aspects of the peritoneal cavity that is an immune-privileged site providing immune competence (Kastelein et al., 2019; Oba et al., 2016; Ono et al., 2006; Sereno et al., 2013; Yap et al., 2021). Following the discovery of EpCAM and the development of triple functional antibodies, a new avenue of research has been conducted. Intraperitoneal immunotherapy with catumaxomab in PC patients has an interesting immunological killing effect on cancer cells in the ascites fluid and diminishes the production of ascites. A randomized phase II/III study in EpCAM-positive cancer patients with ascites requiring puncture for drainage is performed with paracentesis plus catumaxomab or paracentesis alone. Catumaxomab is administered intraperitoneally on days 0, 3, 7 and 10 at doses of 10, 20, 50 and 150 mg, respectively. Paracentesis-free survival

(46 days vs 11, hazard ratio = 0.254: $P < 0.0001$) and median time to next paracentesis (77 vs 13 days; $P < 0.0001$) are significantly longer in the catumaxomab group than the control group. OS has positive trend for the catumaxomab group, especially in patients with gastric cancer ($n = 66$; 71 vs 44 days; $P = 0.0313$). Adverse events associated with catumaxomab (abdominal pain and elevation of liver tests) are frequent manageable and related to the immunological activation (Heiss et al., 2010). Intraperitoneal immunotherapy with catumaxomab associated to chemotherapy administered intravenously was approved in the European Union in April 2009 for the intraperitoneal treatment of patients with malignant ascites. Despite laboratory premises, animal studies, interesting clinical reports and European Medicines Agency authorization the addition of catumaxomab to systemic chemotherapy or given intraperitoneally after surgery in high expression of EpCAM-positive cancer has not entered common clinical practice (Goéré et al., 2014; Kietpeerakool et al., 2019; Knödler et al., 2018).

3.2 Immunoprofiling as a selection for immunotherapy

Patients' selection for immunotherapy is necessary to improve results and optimize costs seems. Retrospective data indicate that immunoprofiling of T cell markers can be prognostic in cancer and useful to identify patients who would benefit from immunotherapy. *Flaherty and colleagues analyze* primary tumor specimens from 89 patients with immunohistochemistry (ICH) for CD3(+), CD4(+), CD8(+) and FOXP3(+) T cells in a phase II trial of nodal staging for colon cancer. Lymphocyte populations were quantified with digital image analysis. They show that longer disease-free survival is associated with higher CD3(+) counts at the invasive margin (IM) ($P = 0.005$), higher CD8(+) counts at the tumor center (TC) and IM ($P = 0.002$), a lower CD4(+)/CD8(+) ratio at the TC + IM ($P = 0.027$) and a higher CD8(+)/FOXP3(+) ratio at the TC + IM ($P = 0.020$). This significance is maintained also after multivariable analysis. They conclude that immunoprofiling in colon cancer after nodal staging has prognostic utility and provide targets for specific immunotherapies (Flaherty et al., 2016).

Based on the same hypothesis Garland-Kledzik and colleagues evaluate immune profiles of patients with colorectal PC who received surgery and HIPEC and their correlation with OS. Eighty tumors are analyzed from 66 patients, including 14 primary and 66 metastatic sites with R0/R1 resection in 44 (66.7%) patients. Known prognostic factors including resection status (HR 1.99, $P = 0.004$) and lymph node status (HR 3.49, $P = 0.002$)

are associated with overall survival. On multivariate analysis, increased CD3/CD4 IM (HR 0.54, $P = 0.03$) ratio is positively associated with improved OS. This study sustains the utility T cells immunoprofiling as prognostic markers in patients with colorectal PC and supports the concept that an enhancement of these cells can offer an effective immunological anti-cancer response (Garland-Kledzik et al., 2021).

3.3 Immunoprofiling after HIPEC

Pro-immunity HIPEC role (drug or hyperthermia related) has not been clearly investigated to date. Some authors show changes in immuno-cells composition in peritoneal lavage or in the blood after HIPEC. Zunino and colleagues report that HIPEC-treated tumor cells induce the activation of tumor-specific T cells and lead to vaccination against tumor cells in mice. They have established that this effect results from the HIPEC-mediated exposure of heat shock protein (HSP) 90 at the plasma membrane. Inhibition or blocking of HSP90, but not HSP70, prevent HIPEC-mediated antitumoral vaccination. This study suggests that HIPEC not only kills tumor cells but also induces an efficient anti-cancer immune response (Zunino et al., 2016).

Franko and colleagues analyze the peritoneal immune cellular profile during cytoreductive surgery and HIPEC and perform flow cytometric analysis of peritoneal fluid collected at laparotomy and during HIPEC at 0, 30, 60, and 90 min. Analysis consisted of standard flow cytometric leukocyte gating, using antibodies for stem cells, B lymphocytes, T-helper, T-suppressor, and natural killer (NK) cells. The proportion of peritoneal NK cells remained stable ($P = 0.655$) throughout perfusion, whereas CD4/CD8 ratio ($P = 0.019$) and granulocyte/lymphocyte ratio ($P = 0.018$) evolved during cytoreduction, with no further change during HIPEC. Two distinct temporal patterns of peritoneal T lymphocytes are evident ("high" and "low" CD4/CD8 ratio groups) and patients maintained their high vs low peritoneal CD4/CD8 ratio status throughout HIPEC duration. They conclude that is mandatory to evaluate peritoneal immunity and the clinical significance of novel peritoneal immune phenotype (Franko et al., 2020).

Fiorentini and colleagues study peripheral blood samples from patients with peritoneal metastases, prior to (day 0) and post-HIPEC procedure (day 30), in order to evaluate the number of $CD3^+$, $CD3^+/CD4^+$ T-Helper, $CD3^+/CD8^+$ cytotoxic T, $CD3^+/CD56^+$ natural killer and $CD19^+$ B lymphocyte and $CD4^+/CD8^+$ T lymphocyte ratios. The total numbers of $CD3^+$, $CD3^+/CD4^+$ T-Helper, $CD3^+/CD8^+$ cytotoxic T, $CD3^+/CD56^+$ natural killer and $CD19^+$ B lymphocytes, and $CD4^+/CD8^+$ lymphocyte ratios increase in all but one patient 30 d following the cytoreductive surgery-HIPEC procedure, and these increases are significant ($P \leq 0.05$) for $CD3^+/CD4^+$ T Helper and $CD3^+/CD8^+$ cytotoxic T lymphocyte numbers. This study suggests that HIPEC has immunomodulating activity, resulting in generalized activation of the adaptive immune response. Moreover, the majority of lymphocyte populations increased following HIPEC and continued to be elevated several weeks following the procedure, consistent with a potential authentic immunomodulating effect rather than a normal inflammatory response (Fiorentini et al., 2020).

Nie and colleagues report that the postoperative administration of HIPEC to patients with advanced adenocarcinoma of the esophagogastric junction significantly increase the portion of CD3+, CD3+CD4+ T cells and reduce the proportion of CD3+CD8+, CD4+CD25+, CD4+PD-1+, CD8+PD-1+ T cells. These results indicate that HIPEC following the surgery can improve the cellular immune function (Nie et al., 2022).

3.4 Immune checkpoint inhibitors

The locoregional treatment of tumors has lost part of its significance due to the fast development of immunotherapy with significant results in melanoma, lung cancer and urological tumors that induces a systemic immune response and the same toxicity.

Immunotherapy is now a mainstay treatment for tumors with DNA defects in mismatch repair proteins and high mutational load, regardless of primary site of origin. Since the peritoneum has the capacity to organize immunological responses, immune checkpoint inhibitors may play a role also in high micro-satellite instability or mismatch repair-deficient metastatic colorectal or ovarian cancer (André et al., 2020; Matulonis et al., 2019; Roth et al., 2021). It has not yet been studied how, when and where to administer checkpoint inhibitors: before surgery with/without HIPEC to promote immune response or later to maintain disease control and prevent relapses. Currently two studies are active in this field of research: one from National University Hospital, Singapore, NCT03172416, where PIPAC is used in combination with nivolumab to monitor if it may have the potential to improve immune activation and response to immune checkpoint inhibition for patients with peritoneal disease (Odendahl et al., 2015) and the second from M.D. Anderson Cancer Center, NCT03508570, where Nivolumab is given intraperitoneally with or without Ipilimumab to treat patients with recurrent or high grade gynecologic cancer with PC.

3.5 Intraperitoneal immunotoxin

An interesting example for a novel treatment approach with an immunotoxin in which the monoclonal antibody MOC31 is covalently linked to Pseudomonas exotoxin A and defined as MOC31PE is provided by the phase I ImmunoPeCa trial. MOC31PE targets cancer-associated epithelial cell adhesion molecules and kills cancer cells by simultaneously causing immune activation by induction of immunogenic cell death (ICD). In this trial Frøysnes and colleagues administer this immunotoxin intraperitoneally one day after surgery and HIPEC, in order to kill EpCAM-positive cancer cells and prevent recurrence. They show that MOC31PE intraperitoneal administration is safe and well tolerated and has low systemic uptake. For this reason, MOC31PE seems ideal for local intraperitoneal treatment. The drug will be further evaluated in an ongoing phase II expansion cohort (Frøysnes et al., 2017).

Recently Thorgersen and colleagues report data on MOC31PE immunotoxin in colorectal patients. They analyzed local and systemic cytokine responses in serum and intraperitoneal fluid samples collected the first three postoperative days (complete resection + HIPEC) with ($n = 12$) or without ($n = 26$) intraperitoneal instillation of MOC31PE, using a wide range, up to 27, of pro- and anti-inflammatory interleukins, interferons, chemokines and other growth factors. They report that activation and intensity of the local and systemic postoperative cytokine response after surgery and HIPEC vary from limited systemic responses to important intraperitoneal responses. Intraperitoneal infusion of MOC31PE result in larger and more intense modifications than radical surgery and HIPEC alone (Thorgersen et al., 2021). The authors reported a significant increase of the values of innate pro-inflammatory cytokines. IL-6, tumor necrosis factor (TNF) and IL-1β are associated to ICD, together with an interesting time response curve for the strong T-cell stimulator interferon-γ and its associated chemokine interferon gamma-induced protein/chemokine ligand 10 were detected. Thorgersen, moreover, underlines a predominately local rather than systemic inflammatory response after surgery and HIPEC, which is enhanced by MOC31PE treatment. The MOC31PE-induced intraperitoneal inflammatory reaction could contribute to improve residual cancer cell killing, however the mechanisms and the toxicity are still not clear. This study is interesting because proposes, with convincing arguments, the use of intraperitoneal immunotherapy, often abandoned in favor of the systemic treatment (Thorgersen et al., 2021).

4. Conclusion

Although PC represents a severe clinical condition with a discouraging prognosis, new results have been achieved thanks to improvements in biological, pharmacological and immunological knowledge. As existing literature suggests, it is time to consider personalized treatment plans for PC patients, associating both clinical and molecular analysis for patient selection. Growing knowledge and insight into the physiology of the immune system may provide a critical step to rise therapeutic weapons to the next level. Future directions may include, novel studies including next generation sequencing and the identification of signaling pathways to improve the best treatment for that specific patient. If there is currently a revolution in which systemic chemotherapy is integrated and replaced with target therapy and agnostic therapy all this cannot fail to recur in the treatment of PC. In other words, while surgical techniques are much improved and standardized, it seems necessary to improve pharmacological and immunological weapons based on the molecular and immunological profiles of patients. The objective is to facilitate a jointing between the conventional clinical and pathologic parameters and new pathways of care at the molecular level to achieve the best therapeutic course for patients.

It is likely that progress made in systemic chemotherapy and check point inhibitors will translate into better response rates for PC and enable radical treatment in a larger number of patients (Roth et al., 2021). Better knowledge on selection criteria will help to identify patients who are likely to profit from intensive surgical approach, whereas patients with an adverse risk profile may undergo only palliative locoregional treatment with PIPAC (Fig. 1). Faced with the impasse in which chemotherapy is often found, the immunological approach for the management of PC appears a new way of emerging research.

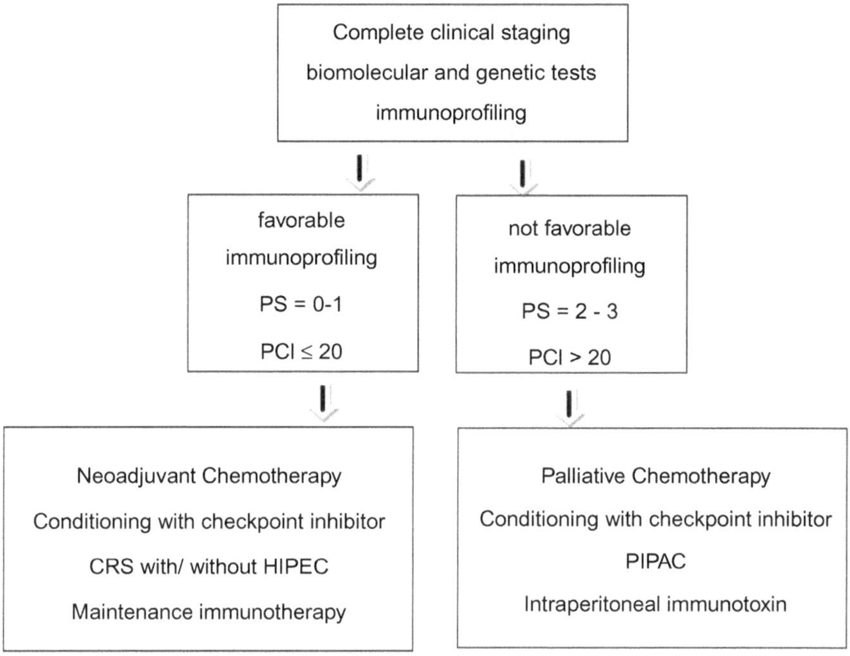

Fig. 1 Flowchart integrating new procedures to improve current results.

Acknowledgment

Dr. Giammaria Fiorentini, MD, is the Former Director of Medical Oncology Unit, Department of Onco-Hematology, Azienda Ospedaliera "Ospedali Riuniti Marche Nord", Pesaro 61122, Italy.

Conflict of interest

The authors do not have any conflict of interest to declare. No funding was received.

References

André, T., Shiu, K.K., Kim, T.W., Jensen, B.V., Jensen, L.H., Punt, C., Smith, D., Garcia-Carbonero, R., Benavides, M., Gibbs, P., de la Fouchardiere, C., Rivera, F., Elez, E., Bendell, J., Le, D.T., Yoshino, T., Van Cutsem, E., Yang, P., Farooqui, M., Marinello, P., KEYNOTE-177 Investigators, 2020. Pembrolizumab in microsatellite-instability-high advanced colorectal cancer. New Eng. J. Med. 38323, 2207–2218. https://doi.org/10.1056/NEJMoa2017699.

Anwar, A., Kasi, A., 2022. Peritoneal Cancer. StatPearls Publishing, Treasure Island, FL. Available from: https://www.ncbi.nlm.nih.gov/books/NBK562138/.

Beniey, M., 2019. Peritoneal metastases from breast cancer: a scoping review. Cureus 118, e5367. https://doi.org/10.7759/cureus.5367.

Breuer, E., Hebeisen, M., Schneider, M.A., Roth, L., Pauli, C., Frischer-Ordu, K., Eden, J., Pache, B., Steffen, T., Hübner, M., Villeneuve, L., Kepenekian, V., Passot, G., Gertsch, P., Gupta, A., Glehen, O., Lehmann, K., 2021. Site of recurrence and survival after surgery for colorectal peritoneal metastasis. J. Natl. Cancer Inst. 1138, 1027–1035. https://doi.org/10.1093/jnci/djab001.

Capobianco, A., Cottone, L., Monno, A., Manfredi, A.A., Rovere-Querini, P., 2017. The peritoneum: healing, immunity, and diseases. J. Pathol. 2432, 137–147. https://doi.org/10.1002/path.4942.

Chen, D., Liu, Z., Liu, W., Fu, M., Jiang, W., Xu, S., Wang, G., Chen, F., Lu, J., Chen, H., Dong, X., Li, G., Chen, G., Zhuo, S., Yan, J., 2021. Predicting postoperative peritoneal metastasis in gastric cancer with serosal invasion using a collagen nomogram. Nat. Commun. 121, 179. https://doi.org/10.1038/s41467-020-20429-0.

Coccolini, F., Gheza, F., Lotti, M., Virzì, S., Iusco, D., Ghermandi, C., Melotti, R., Baiocchi, G., Giulini, S.M., Ansaloni, L., Catena, F., 2013. Peritoneal carcinomatosis. World J. Gastroenterol. 19 (41), 6979–6994. https://doi.org/10.3748/wjg.v19.i41.6979.

de Bree, E., Michelakis, D., Stamatiou, D., Romanos, J., Zoras, O., 2017. Pharmacological principles of intraperitoneal and bidirectional chemotherapy. Pleura Perit. 2 (2), 47–62. https://doi.org/10.1515/pp-2017-0010.

De Simone, M., Vaira, M., Argenziano, M., Berchialla, P., Pisacane, A., Cinquegrana, A., Cavalli, R., Borsano, A., Robella, M., 2020. Pressurized intraperitoneal aerosol chemotherapy (PIPAC) with oxaliplatin, cisplatin, and doxorubicin in patients with peritoneal carcinomatosis: an open-label, single-arm, Phase II Clinical Trial. Biomedicines 8 (5), 102. https://doi.org/10.3390/biomedicines8050102.

Desiderio, J., Chao, J., Melstrom, L., Warner, S., Tozzi, F., Fong, Y., Parisi, A., Woo, Y., 2017. The 30-year experience-A meta-analysis of randomised and high-quality non-randomised studies of hyperthermic intraperitoneal chemotherapy in the treatment of gastric cancer. Eur. J. Cancer Oxford Eng. 1990 (79), 1–14. https://doi.org/10.1016/j.ejca.2017.03.030.

Fiorentini, G., Sarti, D., Patriti, A., Eugeni, E., Guerra, F., Masedu, F., Mackay, A.R., Guadagni, S., 2020. Immune response activation following hyperthermic intraperitoneal chemotherapy for peritoneal metastases: a pilot study. World J. Clin. Oncol. 116, 397–404. https://doi.org/10.5306/wjco.v11.i6.397.

Flaherty, D.C., Lavotshkin, S., Jalas, J.R., Torisu-Itakura, H., Kirchoff, D.D., Sim, M.S., Lee, D.J., Bilchik, A.J., 2016. Prognostic utility of immunoprofiling in colon cancer: results from a prospective, multicenter nodal ultrastaging trial. J. Am. College Surg. 2231, 134–140. https://doi.org/10.1016/j.jamcollsurg.2016.03.003.

Franko, J., Brahmbhatt, R., Tee, M., Raman, S., Ferrel, B., Gorvet, M., Andres, M., 2020. Cellular immunoprofile of peritoneal environment during a HIPEC procedure. Ann. Surg. Oncol. 2713, 5005–5013. https://doi.org/10.1245/s10434-020-08870-3. Epub 2020 Jul 21 32696309.

Frøysnes, I.S., Andersson, Y., Larsen, S.G., Davidson, B., Øien, J.T., Olsen, K.H., Giercksky, K.E., Julsrud, L., Fodstad, Ø., Dueland, S., Flatmark, K., 2017. Novel treatment with intraperitoneal MOC31PE immunotoxin in colorectal peritoneal metastasis: results from the ImmunoPeCa phase 1 trial. Ann. Surg. Oncol. 247, 1916–1922. https://doi.org/10.1245/s10434-017-5814-6.

Garland-Kledzik, M., Uppal, A., Naeini, Y.B., Stern, S., Erali, R., Scholer, A.J., Khader, A.M., Santamaria-Barria, J.A., Cummins-Perry, K., Zhou, Y., Votanopoulos, K.I., Shen, P., Levine, E.A., Bilchik, A.J., 2021. Prognostic impact and utility of immunoprofiling in the selection of patients with colorectal peritoneal carcinomatosis for cytoreductive surgery CRS and heated intraperitoneal chemotherapy HIPEC. J. Gastrointest. Surg. Off. J. Soc. Surg. Alim. Tract 251, 233–240. https://doi.org/10.1007/s11605-020-04886-y.

Goéré, D., Gras-Chaput, N., Aupérin, A., Flament, C., Mariette, C., Glehen, O., Zitvogel, L., Elias, D., 2014. Treatment of gastric peritoneal carcinomatosis by combining complete surgical resection of lesions and intraperitoneal immunotherapy using cetuximab. BMC Cancer 14, 148. https://doi.org/10.1186/1471-2407-14-148.

Hallam, S., Tyler, R., Price, M., Beggs, A., Youssef, H., 2019. Meta-analysis of prognostic factors for patients with colorectal peritoneal metastasis undergoing cytoreductive surgery and heated intraperitoneal chemotherapy. BJS Open 35, 585–594. https://doi.org/10.1002/bjs5.50179.

Heiss, M.M., Ströhlein, M.A., Jäger, M., Kimmig, R., Burges, A., Schoberth, A., Jauch, K.W., Schildberg, F.W., Lindhofer, H., 2005. Immunotherapy of malignant ascites with trifunctional antibodies. Int. J. Cancer 1173, 435–443. https://doi.org/10.1002/ijc.21165.

Heiss, M.M., Murawa, P., Koralewski, P., Kutarska, E., Kolesnik, O.O., Ivanchenko, V.V., Dudnichenko, A.S., Aleknaviciene, B., Razbadauskas, A., Gore, M., Ganea-Motan, E., Ciuleanu, T., Wimberger, P., Schmittel, A., Schmalfeldt, B., Burges, A., Bokemeyer, C., Lindhofer, H., Lahr, A., Parsons, S.L., 2010. The trifunctional antibody catumaxomab for the treatment of malignant ascites due to epithelial cancer: results of a prospective randomized phase II/III trial. Int. J. Cancer 1279, 2209–2221. https://doi.org/10.1002/ijc.25423.

Jaaback, K., Johnson, N., Lawrie, T.A., 2016. Intraperitoneal chemotherapy for the initial management of primary epithelial ovarian cancer. Cochrane Database Syst. Rev. 20161, CD005340. https://doi.org/10.1002/14651858.CD005340.pub4.

Jäger, D., Jäger, E., Knuth, A., 2001. Immune responses to tumour antigens: implications for antigen specific immunotherapy of cancer. J. Clin. Pathol. 549, 669–674. https://doi.org/10.1136/jcp.54.9.669.

Kastelein, A.W., Vos, L., de Jong, K.H., van Baal, J., Nieuwland, R., van Noorden, C., Roovers, J., Lok, C., 2019. Embryology, anatomy, physiology and pathophysiology of the peritoneum and the peritoneal vasculature. Semin. Cell Dev. Biol. 92, 27–36. https://doi.org/10.1016/j.semcdb.2018.09.007s.

Kietpeerakool, C., Rattanakanokchai, S., Jampathong, N., Srisomboon, J., Lumbiganon, P., 2019. Management of Drainage for Malignant Ascites in Gynaecological Cancer. The Cochrane Database of Systematic Reviews. vol. 1212. CD007794 https://doi.org/10.1002/14651858.CD007794.pub3.

Knödler, M., Körfer, J., Kunzmann, V., Trojan, J., Daum, S., Schenk, M., Kullmann, F., Schroll, S., Behringer, D., Stahl, M., Al-Batran, S.E., Hacker, U., Ibach, S., Lindhofer, H., Lordick, F., 2018. Randomised phase II trial to investigate catumaxomab (anti-EpCAM × anti-CD3) for treatment of peritoneal carcinomatosis in patients with gastric cancer. Brit. J. Cancer 1193, 296–302. https://doi.org/10.1038/s41416-018-0150-6.

Kusamura, S., Dominique, E., Baratti, D., Younan, R., Deraco, M., 2008. Drugs, carrier solutions and temperature in hyperthermic intraperitoneal chemotherapy. J. Surg. Oncol. 984, 247–252. https://doi.org/10.1002/jso.21051.

Lemoine, L., Sugarbaker, P., Van der Speeten, K., 2016. Pathophysiology of colorectal peritoneal carcinomatosis: role of the peritoneum. World J. Gastroenterol. 2234, 7692–7707. https://doi.org/10.3748/wjg.v22.i34.7692.

Leung, V., Huo, Y.R., Liauw, W., Morris, D.L., 2017. Oxaliplatin versus mitomycin C for HIPEC in colorectal cancer peritoneal carcinomatosis. Eur. J. Surg. Oncol. 431, 144–149. https://doi.org/10.1016/j.ejso.2016.09.015.

Liu, J., Geng, X., Li, Y., 2016. Milky spots: omental functional units and hotbeds for peritoneal cancer metastasis. Tumour Biol. J. Int. Soc. Oncodev. Biol. Med. 375, 5715–5726. https://doi.org/10.1007/s13277-016-4887-3.

Matulonis, U.A., Shapira-Frommer, R., Santin, A.D., Lisyanskaya, A.S., Pignata, S., Vergote, I., Raspagliesi, F., Sonke, G.S., Birrer, M., Provencher, D.M., Sehouli, J.,

Colombo, N., González-Martín, A., Oaknin, A., Ottevanger, P.B., Rudaitis, V., Katchar, K., Wu, H., Keefe, S., Ruman, J., Ledermann, J.A., 2019. Antitumor activity and safety of pembrolizumab in patients with advanced recurrent ovarian cancer: results from the phase II KEYNOTE-100 study. Ann. Oncol. Off. J. Eur. Soc. Med. Oncol. 307, 1080–1087. https://doi.org/10.1093/annonc/mdz135.

Mulder, W., Ochando, J., Joosten, L., Fayad, Z.A., Netea, M.G., 2019. Therapeutic targeting of trained immunity. Nat. Rev. Drug Discov. 187, 553–566. https://doi.org/10.1038/s41573-019-0025-4.

Mulens-Arias, V., Nicolás-Boluda, A., Pinto, A., Balfourier, A., Carn, F., Silva, A.K.A., Pocard, M., Gazeau, F., 2021. Tumor-selective immune-active mild hyperthermia associated with chemotherapy in colon peritoneal metastasis by photoactivation of fluorouracil-gold nanoparticle complexes. ACS Nano. 23 (152), 3330–3348. https://doi.org/10.1021/acsnano.0c10276. Epub 2021 Feb 2. 33528985.

Nadiradze, G., Horvath, P., Sautkin, Y., Archid, R., Weinreich, F.J., Königsrainer, A., Reymond, M.A., 2019. Overcoming drug resistance by taking advantage of physical principles: pressurized intraperitoneal aerosol chemotherapy PIPAC. Cancers 121, 34. https://doi.org/10.3390/cancers12010034.

Nie, Z., Huang, Y., Li, S., Li, L., 2022. Effect of operation-introducing hyperthermic intraperitoneal chemotherapy on cellular immune function in patients with adenocarcinoma of the esophagogastric junction. Asian J. Surg. 451, 559–560. https://doi.org/10.1016/j.asjsur.2021.09.028. Epub 2021 Oct 20. 34656412.

Oba, M.S., Teramukai, S., Ohashi, Y., Ogawa, K., Maehara, Y., Sakamoto, J., 2016. The efficacy of adjuvant immunochemotherapy with OK-432 after curative resection of gastric cancer: an individual patient data meta-analysis of randomized controlled trials. Gastric Cancer 192, 616–624. https://doi.org/10.1007/s10120-015-0489-9.

Odendahl, K., Solass, W., Demtröder, C., Giger-Pabst, U., Zieren, J., Tempfer, C., Reymond, M.A., 2015. Quality of life of patients with end-stage peritoneal metastasis treated with pressurized intraperitoneal aerosol chemotherapy PIPAC. Eur. J. Surg. Oncol. 4110, 1379–1385. https://doi.org/10.1016/j.ejso.2015.06.001.

Ono, T., Harada, M., Yamada, A., Tanaka, M., Takao, Y., Tanaka, Y., Mine, T., Sakamoto, K., Nakashima, T., Itoh, K., 2006. Antitumor effects of systemic and local immunization with a CTL-directed peptide in combination with a local injection of OK-432. Clin. Cancer Res. 124, 1325–1332. https://doi.org/10.1158/1078-0432.CCR-05-1293.

Prada-Villaverde, A., Esquivel, J., Lowy, A.M., Markman, M., Chua, T., Pelz, J., Baratti, D., Baumgartner, J.M., Berri, R., Bretcha-Boix, P., Deraco, M., Flores-Ayala, G., Glehen, O., Gomez-Portilla, A., González-Moreno, S., Goodman, M., Halkia, E., Kusamura, S., Moller, M., Passot, G., Trout, R., 2014. The American society of peritoneal surface malignancies evaluation of HIPEC with mitomycin C versus oxaliplatin in 539 patients with colon cancer undergoing a complete cytoreductive surgery. J. Surg. Oncol. 1107, 779–785. https://doi.org/10.1002/jso.23728.

Quénet, F., Elias, D., Roca, L., Goéré, D., Ghouti, L., Pocard, M., Facy, O., Arvieux, C., Lorimier, G., Pezet, D., Marchal, F., Loi, V., Meeus, P., Juzyna, B., de Forges, H., Paineau, J., Glehen, O., UNICANCER-GI Group and BIG Renape Group, 2021. Cytoreductive surgery plus hyperthermic intraperitoneal chemotherapy versus cytoreductive surgery alone for colorectal peritoneal metastases PRODIGE 7: a multicentre, randomised, open-label, phase 3 trial. Lancet Oncol. 222, 256–266. https://doi.org/10.1016/S1470-20452030599-4.

Rosenberg, S.A., Lotze, M.T., Yang, J.C., Aebersold, P.M., Linehan, W.M., Seipp, C.A., White, D.E., 1989. Experience with the use of high-dose interleukin-2 in the treatment of 652 cancer patients. Ann. Surg. 2104, 474–485. https://doi.org/10.1097/00000658-198910000-00008.

Rosenberg, S.A., Lotze, M.T., Yang, J.C., Topalian, S.L., Chang, A.E., Schwartzentruber, D.J., Aebersold, P., Leitman, S., Linehan, W.M., Seipp, C.A., 1993. Prospective randomized trial of high-dose interleukin-2 alone or in conjunction with lymphokine-activated killer cells for the treatment of patients with advanced cancer. J. Natl. Cancer Inst. 858, 622–632. https://doi.org/10.1093/jnci/85.8.622.

Roth, L., Russo, L., Ulugoel, S., Freire Dos Santos, R., Breuer, E., Gupta, A., Lehmann, K., 2021. Peritoneal metastasis: current status and treatment options. Cancers 141, 60. https://doi.org/10.3390/cancers14010060.

Ruf, P., Lindhofer, H., 2001. Induction of a long-lasting antitumor immunity by a trifunctional bispecific antibody. Blood 988, 2526–2534. https://doi.org/10.1182/blood.v98.8.2526.

Sadeghi, B., Arvieux, C., Glehen, O., Beaujard, A.C., Rivoire, M., Baulieux, J., Fontaumard, E., Brachet, A., Caillot, J.L., Faure, J.L., Porcheron, J., Peix, J.L., François, Y., Vignal, J., Gilly, F.N., 2000. Peritoneal carcinomatosis from non-gynecologic malignancies: results of the EVOCAPE 1 multicentric prospective study. Cancer 882, 358–363. https://doi.org/10.1002/sici1097-01422000011588:2<358::aid-cncr16>3.0.co;2-o.

Sereno, M., Rodríguez-Esteban, I., Gómez-Raposo, C., Merino, M., López-Gómez, M., Zambrana, F., Casado, E., 2013. Lung cancer and peritoneal carcinomatosis. Oncol. Lett. 63, 705–708. https://doi.org/10.3892/ol.2013.1468.

Solass, W., Kerb, R., Mürdter, T., Giger-Pabst, U., Strumberg, D., Tempfer, C., Zieren, J., Ströhlein, M.A., Siegel, R., Jäger, M., Lindhofer, H., Jauch, K.W., Heiss, M.M., 2009. Induction of anti-tumor immunity by trifunctional antibodies in patients with peritoneal carcinomatosis. J. Exp. Clin. Cancer Res. CR 281, 18. https://doi.org/10.1186/1756-9966-28-18.

Tesniere, A., Schlemmer, F., Boige, V., Kepp, O., Martins, I., Ghiringhelli, F., Aymeric, L., Michaud, M., Apetoh, L., Barault, L., Mendiboure, J., Pignon, J.P., Jooste, V., van Endert, P., Ducreux, M., Zitvogel, L., Piard, F., Kroemer, G., 2010. Immunogenic death of colon cancer cells treated with oxaliplatin. Oncogene 294, 482–491. https://doi.org/10.1038/onc.2009.356.

Thorgersen, E.B., Asvall, J., Frøysnes, I.S., Schjalm, C., Larsen, S.G., Dueland, S., Andersson, Y., Fodstad, Ø., Mollnes, T.E., Flatmark, K., 2021. Increased local inflammatory response to MOC31PE immunotoxin after cytoreductive surgery and hyperthermic intraperitoneal chemotherapy. Ann. Surg. Oncol. 289, 5252–5262. https://doi.org/10.1245/s10434-021-10022-0. Epub 2021 May 21. 34019185. PMC8349350.

van Baal, J., van Noorden, C., Nieuwland, R., Van de Vijver, K.K., Sturk, A., van Driel, W.J., Kenter, G.G., Lok, C., 2018. Development of peritoneal carcinomatosis in epithelial ovarian cancer: a review. J. Histochem. Cytochem. Off. J. Histochem. Soc. 662, 67–83. https://doi.org/10.1369/0022155417742897.

Van der Speeten, K., Stuart, O.A., Sugarbaker, P.H., 2009. Pharmacokinetics and pharmacodynamics of perioperative cancer chemotherapy in peritoneal surface malignancy. Cancer J. Sudbury Mass. 153, 216–224. https://doi.org/10.1097/PPO.0b013e3181a58d95.

van Ruth, S., Verwaal, V.J., Hart, A.A., van Slooten, G.W., Zoetmulder, F.A., 2003. Heat penetration in locally applied hyperthermia in the abdomen during intra-operative hyperthermic intraperitoneal chemotherapy. Anticancer Res. 232B, 1501–1508, 12820416.

Went, P., Vasei, M., Bubendorf, L., Terracciano, L., Tornillo, L., Riede, U., Kononen, J., Simon, R., Sauter, G., Baeuerle, P.A., 2006. Frequent high-level expression of the immunotherapeutic target Ep-CAM in colon, stomach, prostate and lung cancers. Brit. J. Cancer 941, 128–135. https://doi.org/10.1038/sj.bjc.6602924.

Yap, D., Wong, J., Tan, Q.X., Tan, J.W., Chia, C.S., Ong, C.J., 2021. Effect of HIPEC on peritoneal recurrence in peritoneal metastasis treated with cytoreductive surgery: a systematic review. Front. Oncol. 11, 795390. https://doi.org/10.3389/fonc.2021.795390.

Zunino, B., Rubio-Patiño, C., Villa, E., Meynet, O., Proics, E., Cornille, A., Pommier, S., Mondragón, L., Chiche, J., Bereder, J.M., Carles, M., Ricci, J.E., 2016. Hyperthermic intraperitoneal chemotherapy leads to an anticancer immune response via exposure of cell surface heat shock protein 90. Oncogene 352, 261–268. https://doi.org/10.1038/onc.2015.82.

CHAPTER SIX

Omentum: Friend or foe in ovarian cancer immunotherapy?

Ángela Bella[a,b,†,*], Leire Arrizabalaga[a,b,†], Claudia Augusta Di Trani[a,b], Myriam Fernández-Sendin[a,b], Alvaro Teijeira[a,c], Joan Salvador Russo-Cabrera[a,b], Ignacio Melero[a,b,c,d,e,‡], Pedro Berraondo[a,b,c,‡], and Fernando Aranda[a,b,‡,*]

[a]Program of Immunology and Immunotherapy, Cima Universidad de Navarra, Pamplona, Spain
[b]Navarra Institute for Health Research (IdiSNA), Pamplona, Spain
[c]Centro de Investigación Biomédica en Red de Cáncer (CIBERONC), Madrid, Spain
[d]Translational Oncology Group, Program in Solid Tumors, Cima Universidad de Navarra, Pamplona, Spain
[e]Department of Immunology and Immunotherapy, Clinica Universidad de Navarra, Pamplona, Spain
*Corresponding authors: e-mail address: abellacarre@unav.es; faranda@unav.es

Contents

Abstract

Ovarian cancer often spreads out of the ovary before a patient is diagnosed and is the deadliest gynecological malignancy. The aggressiveness of ovarian cancer is determined by the progression in the form of peritoneal carcinomatosis, a stage with a poor prognosis and an untreatable condition in most patients. One of the first tumor nests or the origin of metastasis in the peritoneal cavity is the omentum. The omentum contains immune aggregates, called milky spots, embedded in adipose tissue, which support tumor growth by various mechanisms, including immunosuppressive immune cells and metabolic functions. In this sense, the abundance of blood vessels, omental resident macrophages, and chemokines, among other factors, are known to promote invasiveness, proliferation and resistance to cancer therapies. As a result, surgical practice employed in advanced-stage ovarian cancer almost constantly includes omentectomy. Paradoxically, the omentum is considered the *"abdominal policeman"* that contributes to peritoneal immunity by capturing antigens and pathogens from the peritoneal

† These authors contributed equally to this work.
‡ These authors share senior co-authorship.

International Review of Cell and Molecular Biology, Volume 371
ISSN 1937-6448
https://doi.org/10.1016/bs.ircmb.2022.04.017

cavity and promoting effective immune responses against microbes. Why immunosurveillance against the metastatic tumor does not take place in the omentum? Could omental immune responses be activated with immunotherapeutic interventions? The omentum has largely been ignored in cancer immunology and immunotherapy, and the potential translational implications of this in ovarian cancer are still unclear. Here, we focus on the dual role of the omentum in ovarian cancer: its role in antitumor immune responses *versus* its activities fostering cancer progression.

1. Introduction

In 2020, 313,959 women were diagnosed with ovarian cancer worldwide, and 207,252 women died from the disease (Siegel et al., 2020). Although around 75% of women were alive 5 years after their diagnosis, the Global Cancer Observatory predicts a 47% rise in the incidence of the disease and a 59% increase in the number of deaths each year (Fig. 1)

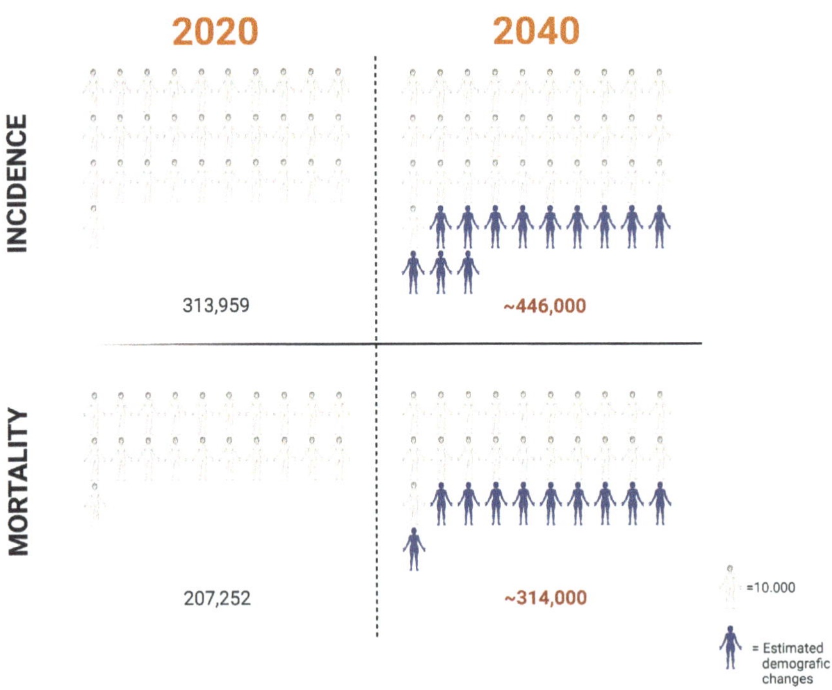

Fig. 1 Ovarian Cancer estimations by Global Cancer Observatory (GCO). GCO is an interactive web-based platform presenting global cancer statistics to inform cancer control and research (https://gco.iarc.fr/). The incidence and mortality of ovarian cancer will increase considerably by 2040.

(Sung et al., 2021). Currently, there is no effective early screening test capable of detecting ovarian cancer (Whitwell et al., 2020). Ovarian cancer is usually diagnosed in advanced stages because the symptoms can be easily confused with less life-threatening digestive conditions such as bloating, constipation, and gas (Arora et al., 2021; Fucikova et al., 2021; Klapp and Galluzzi, 2021). The most frequent ovarian tumors arise from the serosal lining of the ovary, which continues with the serosal lining of the abdominopelvic cavity known as the peritoneum (Michailova and Usunoff, 2006). As a consequence of tumor growth, malignant cells exfoliate and are shed, becoming free and perhaps floating in small quantities of peritoneal fluid. This route of spread of the tumor within the peritoneal cavity gives rise to metastatic implants across the peritoneal cavity (McMullen et al., 2017).

Peritoneal carcinomatosis is an advanced stage, not only in ovarian cancer but also in other gynecological (e.g., endometrial cancer) and gastrointestinal (e.g., colon cancer or stomach cancer) malignancies (Burg et al., 2020; Lambert et al., 2017; Sadeghi et al., 2000; Zhao et al., 2021) in which multiple tumor nodules are thriving in the visceral and parietal peritoneum. Notwithstanding, peritoneal carcinomatosis in ovarian cancer has the highest incidence, with the added concerning clinical fact that at least 40% of women have dissemination to the peritoneum when the first diagnosis of ovarian cancer is made (Bahar and Rokkam, 2021; van Baal et al., 2018; Yeung et al., 2015). Until recently, metastasis in the peritoneum was classified as a terminal stage in oncology and untreatable, with a mean survival of between 2 and 6 months from diagnosis (McMullen et al., 2017). At present, advances in surgery in combination with intraperitoneal chemotherapy have raised the possibility of treating peritoneal carcinomatosis locally (Chia et al., 2016; Gamboa and Winer, 2019; Petroni et al., 2021). However, these treatments have failed to show acceptable efficacy results in terms of improving overall survival (Cavaliere et al., 2009). Peritoneal carcinomatosis at the root of the lethal nature of ovarian cancer and is accompanied by the accumulation of malignant ascites in the peritoneal cavity and the dissemination to other intraperitoneal organs (Cao et al., 2021; Ozols et al., 2004). Despite the range of organs located in the peritoneal cavity, one in particular stands out regarding the progression of peritoneal carcinomatosis, the omentum (Platell et al., 2000).

Metastasis to the omentum is most frequently the beginning of peritoneal carcinomatosis, particularly in cancers arising from the ovaries (Dawson et al., 2021; van Baal et al., 2018). The omentum is a peritoneum-coated membrane of abdominal fat derived from mesothelial cells nestling on the surface of the intra-peritoneal organs. In humans, there are two omenta: (i) the greater omentum that extends like an apron over the organs of the digestive system; and (ii) the lesser omentum, which connects the lesser curvature of the stomach and the proximal duodenum to the liver. The omentum plays a key role in fat storage, neovascularization and tissue regeneration (Di Nicola, 2019; Platell et al., 2000). Further, the omentum was described as "the policeman of the abdomen" for being responsible for the peritoneal immunity through functional aggregates of different immune cells called "milky spots" (Beelen, 1991, 1992; Brestoff et al., 2015; Lee et al., 2015; Mebius, 2009; Meza-Perez and Randall, 2017; Rangel-Moreno et al., 2009).

To remove the omentum is considered critical in advanced ovarian cancer when debulking or cytoreduction surgeries are performed (Di Nicola, 2019). In ovarian cancer, surgeries intend to remove as many peritoneal implants as possible in a debulking approach, so there is less cancer left to treat with subsequent anti-cancer treatments. When much of the peritoneal carcinomatosis is removed surgically, there is a greater chance of getting rid of the leftover ovarian cancer cells. Previously, tumor cells exfoliate from the ovary nest and grow in the omentum because it has many tiny blood vessels and is rich in pro-tumoral cells and molecules that constitute a favorable environment for cancer progression (Fig. 2) (Brestoff et al., 2015; Gerber et al., 2006). Moreover, as the omentum is composed mainly of adipocytes storing triglycerides, ovarian cancer cells use those lipids as nutrients to further develop the peritoneal carcinomatosis (Nieman et al., 2011). However, in contradiction to this pro-tumoral role in ovarian cancer, the omentum is considered the keeper of the peritoneal cavity in normal conditions (Di Nicola, 2019; Liu et al., 2021a). Accordingly, the omentum filters out putative pathogens in the peritoneum, which are eliminated by immune responses (Liu et al., 2021b). However, despite the omentum being an organized tissue of immune cells, it has been shown that as a defense mechanism, it does not function in cancer (Liu et al., 2021a). When tumor cells migrate to the omentum, instead of activating the immune effector mechanisms for their elimination, some degree of tolerance seems to occur that actually favors the progression of peritoneal carcinomatosis. Multiple factors are at play, such as the lack of dendritic cells ($CD8^+CD11b^-$) specialized to *cross-present* tumor antigens to resident lymphocytes (Carlow et al., 2009; Elpek et al., 2011) or the pro-tumoral role of a macrophage $CD163^+$ $Tim4^+$ subset in the omentum (Etzerodt et al., 2020). Could antitumor immune responses be activated or reactivated in the omentum?

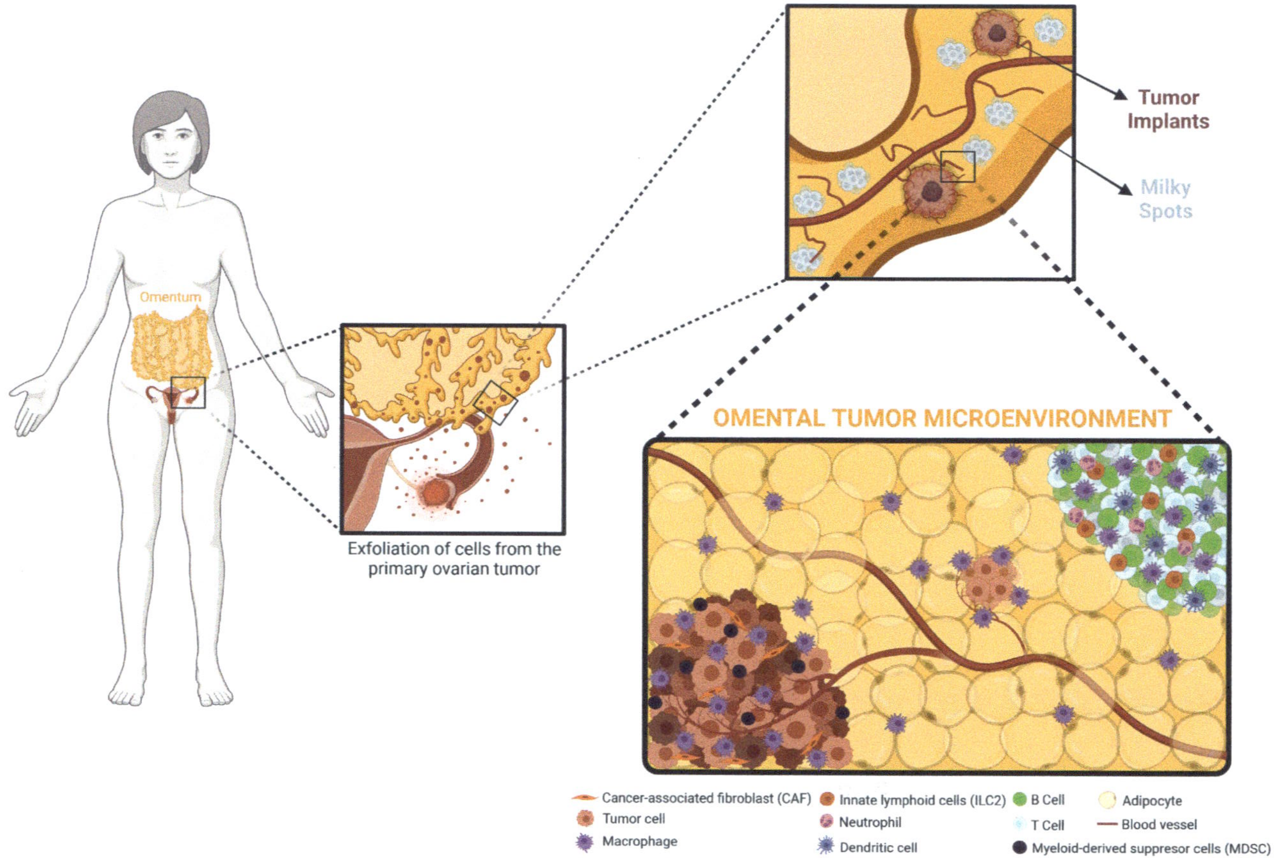

Fig. 2 Omental tumor microenvironment in ovarian cancer. Primary ovarian cancer cells are prone to be exfoliated to the peritoneum. The first place where malignant cells nest is in the omentum. Although the omentum is considered an immune tissue, the immune activation to eliminate the nesting of tumor cells does not occur, and thus cancer progresses. In addition, the high omental vascularization and the presence of pro-tumor immune cells in the omentum, such as macrophages or myeloid suppressor cells, hastens progression in advanced-stage ovarian cancer.

In the next sections, we review current knowledge regarding the immune antitumor activities and protumoral functions of the omentum to understand better the role of this peculiar tissue in metastatic ovarian cancer.

2. Milky spots and their immunological role

Although the omentum is primarily a fatty tissue, it contains clusters of leukocytes termed "*milky spots*" or "*taches laiteuses*" as they were originally termed. The existence of milky spots was discovered in 1874 by the French anatomist Ranvier, L. when he localized white spots of immune cells in the omentum of rabbits (Platell et al., 2000). These immune clusters (Fig. 2) consisting of T and B lymphocytes, innate lymphoid cells (ILC2), neutrophils, macrophages, and dendritic cells that collectively contribute to peritoneal immunity by capturing antigens, particulates, and pathogens present in the peritoneal cavity and promoting a variety of immune responses, including inflammation, tolerance, and fibrosis (Beelen, 1991, 1992; Brestoff et al., 2015; Lee et al., 2015; Liu et al., 2021a; Mebius, 2009; Meza-Perez and Randall, 2017; Rangel-Moreno et al., 2009). Anatomically, in humans, there are two omenta (larger and lesser omenta linked to the stomach). While the size and shape of the omentum vary depending on mammalian species, milky spots speckle the entire tissue and work as a filter for the surrounding intracavitary fluid (Shimotsuma et al., 1993). The fluid around the abdominal organs circulates through the milky spots, and these may collect cells, antigens, and pathogens (bacteria, virus, etc.) (Beelen, 1992). In this regard, milky spots regulate immunity by releasing cytokines and chemokines (LT-β and TNF-α, as well as XCL12, CCL21, CCL19, and CXCL13), inducing antigen tolerance, and beginning the process of fibrosis, among others, all of which clearly show the omentum's role as a primary immune organ (Mebius, 2009; Rangel-Moreno et al., 2009; Shimotsuma et al., 1993).

Unfortunately, in the case of tumor cells, the omentum does not behave as a peritoneal protector. In this scenario, the omentum provides tolerance instead of immunity to respond to metastatic tumor cells (van Baal et al., 2018; Yeung et al., 2015). In the case of ovarian cancer, the circulating fluids bring exfoliated ovarian cancer cells to the milky spots, where these tumor cells become entrapped, promoting successful omental metastases (Hagiwara et al., 1993; van Baal et al., 2018). However, the immune response of the omentum has not been widely explored for its antitumor potential in the metastasis of ovarian cancer. Some preclinical studies point to immune activation or antigen targeting strategies in the peritoneum that can make the omentum an ally in the response against peritoneal carcinomatosis and thus the primary cancer tumor (Carlow et al., 2009; Sedlacek et al., 2013; Sorensen et al., 2009). These findings suggest that omentectomy could be an inappropriate surgical practice in some particular cases, eliminating the potential of resident immune cells. In this regard, the activity of these omental immune cells could perhaps be modulated by immunotherapeutic strategies, and thus, ovarian cancer progression in the peritoneal cavity could be controlled. In relation to this, Sedlacek et al. observed that the peritoneal cavity has a peculiar environment capable of effecting potent specific and nonspecific antitumor immune mechanisms (Sedlacek et al., 2013). In an attempt to determine whether the omentum is capable of promoting antitumor responses, mice were immunized *i.p.,* with irradiated cells from a colon tumor cell line (MC38). They observed this immunization was potentially effective only when the omentum was present by the release of "*immunogenic cell death*" factors or alarmins (e.g., TNF-α, IL-6) (Fucikova et al., 2020), ultimately inducing a $CD4^+$ and $CD8^+$ T cell-specific antitumor responses (Paijens et al., 2021). In addition, cellular depletion of immune subsets revealed that omental natural killer cells are unique, capable of preventing the growth of both specific and unrelated tumors and mediate protection in peritoneal carcinomatosis mouse models. Similarly, the same group has shown that specific immunization strategies targeting milky spots elicit a potent antitumor immune response in an ovarian cancer model (ID8) (Sorensen et al., 2009). They showed that omental immune cells were more effective in generating an effective adaptive immune response against tumor cells as compared to immune cells in the spleen or in peritoneal lavages. Finally, the authors concluded that the omentum could respond strongly to antigens and do so more effectively than other peritoneal immune organs to generate peritoneal carcinomatosis immune responses.

With regard to intracavitary adoptive cell transfer interventions, Carlow et al. infused lymphocytes by the intraperitoneal route and demonstrated how these immune cells were activated in the omentum by resident dendritic cells *cross-presenting antigens* (Carlow et al., 2009). These authors showed that the intraperitoneal infusion of lymphocytes led to their rapid accumulation in the omentum, with the highest concentration of infused lymphocytes capable of responding to antigens being reached after 1–2 h but with accumulation still continuing up to 20 h later. This immunological feature could be of great interest toward the implementation of chimeric antigen receptor (CAR)-T intracavitary peritoneal carcinomatosis interventions.

The omentum and immunotherapy intraperitoneal interventions have not been explored to date. However, experimental evidence exists to support that the omentum may become an effective antitumor immune zone if appropriate immunotherapy strategies are devised.

3. Tumor metastasis and ovarian peritoneal carcinomatosis origin

Ovarian cancer is, in most cases, diagnosed at an advanced stage or stage III when the neoplasia has spread to distant peritoneal organs (Arora et al., 2021). In particular, the spread of ovarian cancer to the omentum is associated with a poor prognosis and ultimately results in very aggressive peritoneal carcinomatosis. Thus, the omentum plays a critical role and has shown itself to be the preferred nesting tissue for ovarian tumor cells exfoliated from the ovary (Iwagoi et al., 2021). Experimental mouse models support this notion, as it has been shown that ovarian cancer cells (murine, ID8; human, SKOV3, CaOV3, HeyA8) colonize the omentum within the first 6h following intraperitoneal inoculation (Sorensen et al., 2009).

Some immune cells present in the milky spots, including neutrophils and macrophages, have been shown to support the metastasis of ovarian cancer to the omentum. Further, many studies have presented the macrophages of the peritoneum and omentum as the preferred allies of ovarian cancer malignant cells (Hensler et al., 2020; Vitale et al., 2019). Using mouse models, Etzerodt et al. demonstrated that a specific subset of tissue-resident macrophages in the omentum which are characterized by the high expression of CD163 and Tim4, are required for the metastatic spread of ovarian cancer cells (Etzerodt et al., 2020). These authors have also shown that selective depletion of $CD163^{+}$ $Tim4^{+}$ macrophages prevented the development of omental metastasis, whereas the depletion of monocyte-derived tumor-associated macrophages had no impact. Similar work has reported that omental macrophages promote ovarian cancer colonization of the omentum through the secretion of chemokines. Krishnan et al. hypothesized that CCL6 in mouse ovarian cancer models, and CCL23, the human ortholog of CCL6, are the primary mediators of cancer colonization of the omentum (Krishnan et al., 2020). Using genetic and pharmacologic inhibition of the CCL6/CCL23 receptor and chemokine receptor 1 (CCR1), they observed a marked reduction of cancer cell migration from the ovary to the omentum. In addition to this, different stromal cell types are involved in ovarian cancer metastasis and immune evasion mechanisms (Gregg, 2021). Among them, endothelial cells, cancer-associated fibroblast, macrophages and adipocytes play essential roles by cytokines and growth factors released into the tumor microenvironment (Yeung et al., 2015). For instance, immune cells suppression is induced by VEGF, IL-6, IL-10, LIF, and TGF-β (Worzfeld et al., 2017).

Likewise, neutrophils promote the premetastatic omentum in ovarian cancer. Lee et al. observed an accumulation of neutrophils in the premetastatic niche of omenta which were found to facilitate the implantation of ovarian cancer cells due to chromatin webs called NETs (neutrophil extracellular traps) that bind ovarian cancer cells inducing metastasis (Lee et al., 2019; Peng et al., 2021). Previous studies have detected neutrophil chemotactic factors at high levels in body fluids of women with ovarian cancer, including IL-8, monocyte chemoattractant protein-1 (MCP-1) (Negus et al., 1995), and granulocyte colony-stimulating factor (G-CSF) (Gorelik et al., 2005). In addition, local neutrophil extracellular trap (NETs) formation has been incriminated to play a pro-tumor role in the premetastatic omental niche involved in the successful implantation of primary ovarian cancer cells. Thus, Lee et al. have demonstrated that the pharmacological blockade of NET formation substantially decreased omental metastasis. This inhibition of NET formation consists of the treatment with peptidyl arginine deiminase 4 (PAD4) inhibitor or GSK484, a small molecule inhibitor (Lewis et al., 2015).

Interestingly, although the omentum collects exfoliated ovarian cancer cells, tumor cells circulating in the blood (hematogenous metastasis) also show tropism for the omentum. The group of Sunila Pradeep demonstrated that high levels of receptor tyrosine-protein kinase ErbB-3, also known as HER3 (human epidermal growth factor receptor 3), in ovarian cancer cells and neuregulin 1 (NRG1) in the omentum facilitate tumor cell colonization and eventual growth (Chen et al., 2020). They demonstrated that the therapeutic disruption of the Erb3/NRG1 signaling axis in mice reduces hematogenous metastasis to the omentum in ovarian cancer models. This strategy has been tested in therapeutic combinations such as the use of bevacizumab, nintedanib, and pazopanib in women responding to front-line and/or recurrent platinum-based chemotherapy (Aghajanian et al., 2012; Daniele et al., 2021; Perren et al., 2011), cediranib in women with platinum-based chemotherapy for recurrent ovarian cancer (Ledermann et al., 2018), and paclitaxel (Francis et al., 1995; Markman et al., 1992). In this regard, the human monoclonal anti-ErbB3 antibody, seribantumab (MM-121), has been studied in preclinical ovarian cancer PDX models (Odintsov et al., 2021) as well in a clinical trial (NCT01447706) (Liu et al., 2016) that did not find evidence for efficacy.

In advanced-metastatic ovarian cancer, the omental fibroblast subpopulation, defined as cancer-associated fibroblasts (CAFs), supports and promotes omental tumor progression (Zhang et al., 2011). In fact, the first nested ovarian cancer cells in the omentum stimulate the proliferation of omental CAFs and induce mesothelial–mesenchymal transition (MMT) of peritoneal mesothelial cells *via* transforming growth factor β1 (TGF-β1) (Cai et al., 2012; Tang et al., 2020). Further, CAFs secrete a number of important protumoral molecules, including angiogenetic factors, prolymphangiogenic factors, and metastasis-promoting factors such as vascular endothelial growth factor (VEGF). Omental expression of VEGF-A, VEGF-D, VEGF receptor-1 in patients with ovarian cancer was significantly higher in omental metastases than the related primary ovarian tumors (Sopo et al., 2019). In this sense, VEGF promotes omental angiogenesis and, in consequence, the promotion of ovarian cancer cell implantation and tumor growth in the milky spots. Furthermore, an elevated level of TGF-β1 promotes an immunosuppressive environment recruiting regulatory T-cells (Tregs) and impairing cytotoxic T and NK cells functions (Batlle and Massague, 2019; Ghiringhelli et al., 2005). Increased Treg/$CD8^+$ ratio was found in the omentum of metastasis bearing mice and TGF-β1 neutralization invert the ratio by significantly increasing $CD8^+$ T cells over Tregs (Roane et al., 2021). Moreover, TGF-β pathway was found activated in omental metastases compared to primary disease sites (Yamamura et al., 2012) and TGF-β1 has been found in malignant ascites-derived exosomes (Wei et al., 2017). In summary, these data highlight the potential therapeutical benefit of blocking TGF-β signaling in peritoneal metastasis.

Overall, it is well established that ovarian cancer cell migration and nesting to the omentum are not passive and haphazard phenomena. Further, the omentum provides the ideal conditions for converting the primary ovarian cancer tumor to an advanced stage. Immune cells residing in the milky spots are polarized toward an undesirable immunosuppressive state with the increased presence of regulatory T cells, myeloid suppressor cells, M2 macrophages, etc. Several intracavitary or systemic immunomodulatory therapy strategies can lead the omentum to become hostile for cancer progression and a driver of antitumor immunity.

4. Concluding remarks

The main question that arises is whether this omental region can be modulated with immunotherapeutic strategies. Previous work suggests this may be the case, taking advantage of the site where the tumor cells are concentrated and the fact that regional immune cells co-exist in this scenario but in a yet ill-defined tolerogenic state. More studies are needed to determine when the omentum can become an ally or an enemy in ovarian cancer management and if it can be a target, especially with intraperitoneal immunotherapeutic strategies that offer more opportunities in the treatment of peritoneal carcinomatosis perhaps broadening the therapeutic window of existing agents as a result of a reduction of systemic exposure.

In conclusion, the omentum is clearly a foe in the evolution of ovarian cancer patients but endowed with immune features that perhaps might one day permit to turn this devil into an angel.

Acknowledgments

We are grateful to Dr. Paul Miller for English editing. The figures were created by BioRender website platform. This work was supported by Instituto de Salud Carlos III (PI20/00002, PI19/01128) cofinanced by Fondos FEDER "A way to make Europe" and Joint Translational Call for Proposals 2015 (JTC 2015), TRANSCAN456 2 (code: TRS-2016-00000371), Gobierno de Navarra Proyecto LINTERNA Ref.: 0011-1411-2020-000075. This project has received funding from the European Union's Horizon 2020 research and innovation program under the Marie Skłodowska-Curie grant agreement No 765394. F.A. receives a Miguel Servet I (CP19/00114) contract from ISCIII (Instituto de Salud Carlos III) co-financed by FSE (Fondo Social Europeo) "Investing in your future." A.B. is a recipient of a PFIS fellowship from ISCIII (FI20/00058).

References

Aghajanian, C., Blank, S.V., Goff, B.A., Judson, P.L., Teneriello, M.G., Husain, A., et al., 2012. OCEANS: a randomized, double-blind, placebo-controlled phase III trial of chemotherapy with or without bevacizumab in patients with platinum-sensitive recurrent epithelial ovarian, primary peritoneal, or fallopian tube cancer. J. Clin. Oncol. 30 (17), 2039–2045.

Arora, T., Mullangi, S., Lekkala, M.R., 2021. Ovarian Cancer. StatPearls, Treasure Island (FL).

Bahar, S.G., Rokkam, V.R., 2021. Omentum Tumors. StatPearls, Treasure Island (FL).

Batlle, E., Massague, J., 2019. Transforming growth factor-beta signaling in immunity and cancer. Immunity 50 (4), 924–940.

Beelen, R.H., 1991. The greater omentum: physiology and immunological concepts. Neth. J. Surg. 43 (5), 145–149.

Beelen, R.H., 1992. Role of omental milky spots in the local immune response. Lancet 339 (8794), 689.
Brestoff, J.R., Kim, B.S., Saenz, S.A., Stine, R.R., Monticelli, L.A., Sonnenberg, G.F., et al., 2015. Group 2 innate lymphoid cells promote beiging of white adipose tissue and limit obesity. Nature 519 (7542), 242–246.
Burg, L., Timmermans, M., van der Aa, M., Boll, D., Rovers, K., de Hingh, I., et al., 2020. Incidence and predictors of peritoneal metastases of gynecological origin: a population-based study in the Netherlands. J. Gynecol. Oncol. 31 (5), e58.
Cai, J., Tang, H., Xu, L., Wang, X., Yang, C., Ruan, S., et al., 2012. Fibroblasts in omentum activated by tumor cells promote ovarian cancer growth, adhesion and invasiveness. Carcinogenesis 33 (1), 20–29.
Cao, K., Zhang, G., Zhang, X., Yang, M., Wang, Y., He, M., et al., 2021. Stromal infiltrating mast cells identify immunoevasive subtype high-grade serous ovarian cancer with poor prognosis and inferior immunotherapeutic response. Onco. Targets. Ther. 10 (1), 1969075.
Carlow, D.A., Gold, M.R., Ziltener, H.J., 2009. Lymphocytes in the peritoneum home to the omentum and are activated by resident dendritic cells. J. Immunol. 183 (2), 1155–1165.
Cavaliere, F., Giannarelli, D., Valle, M., Federici, O., Liotta, G., Perri, P., et al., 2009. Peritoneal carcinomatosis from ovarian epithelial primary: combined aggressive treatment. In Vivo 23 (3), 441–446.
Chen, C., Gupta, P., Parashar, D., Nair, G.G., George, J., Geethadevi, A., et al., 2020. ERBB3-induced furin promotes the progression and metastasis of ovarian cancer via the IGF1R/STAT3 signaling axis. Oncogene 39 (14), 2921–2933.
Chia, C.S., You, B., Decullier, E., Vaudoyer, D., Lorimier, G., Abboud, K., et al., 2016. Patients with peritoneal carcinomatosis from gastric cancer treated with cytoreductive surgery and hyperthermic intraperitoneal chemotherapy: is cure a possibility? Ann. Surg. Oncol. 23 (6), 1971–1979.
Daniele, G., Raspagliesi, F., Scambia, G., Pisano, C., Colombo, N., Frezzini, S., et al., 2021. Bevacizumab, carboplatin, and paclitaxel in the first line treatment of advanced ovarian cancer patients: the phase IV MITO-16A/MaNGO-OV2A study. Int. J. Gynecol. Cancer 31 (6), 875–882.
Dawson, M.R., Xuan, B., Hsu, J., Ghosh, D., 2021. Force balancing ACT-IN the tumor microenvironment: cytoskeletal modifications in cancer and stromal cells to promote malignancy. Int. Rev. Cell Mol. Biol. 360, 1–31.
Di Nicola, V., 2019. Omentum a powerful biological source in regenerative surgery. Regen Ther. 11, 182–191.
Elpek, K.G., Bellemare-Pelletier, A., Malhotra, D., Reynoso, E.D., Veronika, L.-K., DeKruyff, R.H., Turley, S.J., et al., 2011. Lymphoid organ-resident dendritic cells exhibit unique transcriptional fingerprints based on subset and site. PLoS One 6 (8), e23921.
Etzerodt, A., Moulin, M., Doktor, T.K., Delfini, M., Mossadegh-Keller, N., Bajenoff, M., et al., 2020. Tissue-resident macrophages in omentum promote metastatic spread of ovarian cancer. J. Exp. Med. 217 (4).
Francis, P., Rowinsky, E., Schneider, J., Hakes, T., Hoskins, W., Markman, M., 1995. Phase I feasibility and pharmacologic study of weekly intraperitoneal paclitaxel: a gynecologic oncology group pilot study. J. Clin. Oncol. 13 (12), 2961–2967.
Fucikova, J., Kepp, O., Kasikova, L., Petroni, G., Yamazaki, T., Liu, P., et al., 2020. Detection of immunogenic cell death and its relevance for cancer therapy. Cell Death Dis. 11 (11), 1013.
Fucikova, J., Coosemans, A., Orsulic, S., Cibula, D., Vergote, I., Galluzzi, L., et al., 2021. Immunological configuration of ovarian carcinoma: features and impact on disease outcome. J. Immunother. Cancer 9 (10).

Gamboa, A.C., Winer, J.H., 2019. Cytoreductive surgery and hyperthermic intraperitoneal chemotherapy for gastric cancer. Cancers (Basel) 11 (11).

Gerber, S.A., Rybalko, V.Y., Bigelow, C.E., Lugade, A.A., Foster, T.H., Frelinger, J.G., et al., 2006. Preferential attachment of peritoneal tumor metastases to omental immune aggregates and possible role of a unique vascular microenvironment in metastatic survival and growth. Am. J. Pathol. 169 (5), 1739–1752.

Ghiringhelli, F., Menard, C., Terme, M., Flament, C., Taieb, J., Chaput, N., et al., 2005. CD4+CD25+ regulatory T cells inhibit natural killer cell functions in a transforming growth factor-beta-dependent manner. J. Exp. Med. 202 (8), 1075–1085.

Gorelik, E., Landsittel, D.P., Marrangoni, A.M., Modugno, F., Velikokhatnaya, L., Winans, M.T., et al., 2005. Multiplexed immunobead-based cytokine profiling for early detection of ovarian cancer. Cancer Epidemiol. Biomarkers Prev. 14 (4), 981–987.

Gregg, R.K., 2021. Implications of microgravity-induced cell signaling alterations upon cancer cell growth, invasiveness, metastatic potential, and control by host immunity. Int. Rev. Cell Mol. Biol. 361, 107–164.

Hagiwara, A., Takahashi, T., Sawai, K., Taniguchi, H., Shimotsuma, M., Okano, S., et al., 1993. Milky spots as the implantation site for malignant cells in peritoneal dissemination in mice. Cancer Res. 53 (3), 687–692.

Hensler, M., Kasikova, L., Fiser, K., Rakova, J., Skapa, P., Laco, J., et al., 2020. M2-like macrophages dictate clinically relevant immunosuppression in metastatic ovarian cancer. J. Immunother. Cancer 8 (2).

Iwagoi, Y., Motohara, T., Hwang, S., Fujimoto, K., Ikeda, T., Katabuchi, H., 2021. Omental metastasis as a predictive risk factor for unfavorable prognosis in patients with stage III-IV epithelial ovarian cancer. Int. J. Clin. Oncol. 26 (5), 995–1004.

Klapp, V., Galluzzi, L., 2021. Multimodal immunostimulation to control BRCA1-defective ovarian carcinoma. Trends. Cancer.

Krishnan, V., Tallapragada, S., Schaar, B., Kamat, K., Chanana, A.M., Zhang, Y., et al., 2020. Omental macrophages secrete chemokine ligands that promote ovarian cancer colonization of the omentum via CCR1. Commun. Biol. 3 (1), 524.

Lambert, A.W., Pattabiraman, D.R., Weinberg, R.A., 2017. Emerging biological principles of metastasis. Cell 168 (4), 670–691.

Ledermann, J.A., Raja, F.A., Fotopoulou, C., Gonzalez-Martin, A., Colombo, N., Sessa, C., et al., 2018. Newly diagnosed and relapsed epithelial ovarian carcinoma: ESMO clinical practice guidelines for diagnosis, treatment and follow-up. Ann. Oncol. 29 (Suppl 4), (iv259).

Lee, M.W., Odegaard, J.I., Mukundan, L., Qiu, Y., Molofsky, A.B., Nussbaum, J.C., et al., 2015. Activated type 2 innate lymphoid cells regulate beige fat biogenesis. Cell 160 (1–2), 74–87.

Lee, W., Ko, S.Y., Mohamed, M.S., Kenny, H.A., Lengyel, E., Naora, H., 2019. Neutrophils facilitate ovarian cancer premetastatic niche formation in the omentum. J. Exp. Med. 216 (1), 176–194.

Lewis, H.D., Liddle, J., Coote, J.E., Atkinson, S.J., Barker, M.D., Bax, B.D., et al., 2015. Inhibition of PAD4 activity is sufficient to disrupt mouse and human NET formation. Nat. Chem. Biol. 11 (3), 189–191.

Liu, J.F., Ray-Coquard, I., Selle, F., Poveda, A.M., Cibula, D., Hirte, H., et al., 2016. Randomized phase II trial of Seribantumab in combination with paclitaxel in patients with advanced platinum-resistant or -refractory ovarian cancer. J. Clin. Oncol. 34 (36), 4345–4353.

Liu, M., Silva-Sanchez, A., Randall, T.D., Meza-Perez, S., 2021a. Specialized immune responses in the peritoneal cavity and omentum. J. Leukoc. Biol. 109 (4), 717–729.

Liu, Y., Hu, J.N., Luo, N., Zhao, J., Liu, S.C., Ma, T., et al., 2021b. The essential involvement of the Omentum in the peritoneal defensive mechanisms during intra-abdominal sepsis. Front. Immunol. 12, 631609.

Markman, M., Rowinsky, E., Hakes, T., Reichman, B., Jones, W., Lewis, J.L., Jr., et al., 1992. Phase I trial of intraperitoneal taxol: a Gynecoloic oncology group study. J. Clin. Oncol. 10 (9), 1485–1491.

McMullen, J.R.W., Selleck, M., Wall, N.R., Senthil, M., 2017. Peritoneal carcinomatosis: limits of diagnosis and the case for liquid biopsy. Oncotarget 8 (26), 43481–43490.

Mebius, R.E., 2009. Lymphoid organs for peritoneal cavity immune response: milky spots. Immunity 30 (5), 670–672.

Meza-Perez, S., Randall, T.D., 2017. Immunological functions of the Omentum. Trends Immunol. 38 (7), 526–536.

Michailova, K.N., Usunoff, K.G., 2006. Serosal membranes (pleura, pericardium, peritoneum). Normal structure, development and experimental pathology. Adv. Anat. Embryol. Cell Biol. 183 (i-vii), 1–144 (back cover).

Negus, R.P., Stamp, G.W., Relf, M.G., Burke, F., Malik, S.T., Bernasconi, S., et al., 1995. The detection and localization of monocyte chemoattractant protein-1 (MCP-1) in human ovarian cancer. J. Clin. Invest. 95 (5), 2391–2396.

Nieman, K.M., Kenny, H.A., Penicka, C.V., Ladanyi, A., Buell-Gutbrod, R., Zillhardt, M.R., et al., 2011. Adipocytes promote ovarian cancer metastasis and provide energy for rapid tumor growth. Nat. Med. 17 (11), 1498–1503.

Odintsov, I., Lui, A.J.W., Sisso, W.J., Gladstone, E., Liu, Z., Delasos, L., et al., 2021. The anti-HER3 mAb Seribantumab effectively inhibits growth of patient-derived and isogenic cell line and xenograft models with oncogenic NRG1 fusions. Clin. Cancer Res. 27 (11), 3154–3166.

Ozols, R.F., Bookman, M.A., Connolly, D.C., Daly, M.B., Godwin, A.K., Schilder, R.J., et al., 2004. Focus on epithelial ovarian cancer. Cancer Cell 5 (1), 19–24.

Paijens, S.T., Vledder, A., Loiero, D., Duiker, E.W., Bart, J., Hendriks, A.M., et al., 2021. Prognostic image-based quantification of CD8CD103 T cell subsets in high-grade serous ovarian cancer patients. Onco. Targets. Ther. 10 (1), 1935104.

Peng, Z., Liu, C., Victor, A.R., Cao, D.Y., Veiras, L.C., Bernstein, E.A., et al., 2021. Tumors exploit CXCR4(hi)CD62L(lo) aged neutrophils to facilitate metastatic spread. Onco. Targets. Ther. 10 (1), 1870811.

Perren, T.J., Swart, A.M., Pfisterer, J., Ledermann, J.A., Pujade-Lauraine, E., Kristensen, G., et al., 2011. A phase 3 trial of bevacizumab in ovarian cancer. N. Engl. J. Med. 365 (26), 2484–2496.

Petroni, G., Buque, A., Zitvogel, L., Kroemer, G., Galluzzi, L., 2021. Immunomodulation by targeted anticancer agents. Cancer Cell 39 (3), 310–345.

Platell, C., Cooper, D., Papadimitriou, J.M., Hall, J.C., 2000. The omentum. World J. Gastroenterol. 6 (2), 169–176.

Rangel-Moreno, J., Moyron-Quiroz, J.E., Carragher, D.M., Kusser, K., Hartson, L., Moquin, A., et al., 2009. Omental milky spots develop in the absence of lymphoid tissue-inducer cells and support B and T cell responses to peritoneal antigens. Immunity 30 (5), 731–743.

Roane, B.M., Meza-Perez, S., Katre, A.A., Goldsberry, W.N., Randall, T.D., Norian, L.A., et al., 2021. Neutralization of TGFbeta improves tumor immunity and reduces tumor progression in ovarian carcinoma. Mol. Cancer Ther. 20 (3), 602–611.

Sadeghi, B., Arvieux, C., Glehen, O., Beaujard, A.C., Rivoire, M., Baulieux, J., et al., 2000. Peritoneal carcinomatosis from non-gynecologic malignancies: results of the EVOCAPE 1 multicentric prospective study. Cancer 88 (2), 358–363.

Sedlacek, A.L., Gerber, S.A., Randall, T.D., van Rooijen, N., Frelinger, J.G., Lord, E.M., 2013. Generation of a dual-functioning antitumor immune response in the peritoneal cavity. Am. J. Pathol. 183 (4), 1318–1328.

Shimotsuma, M., Shields, J.W., Simpson-Morgan, M.W., Sakuyama, A., Shirasu, M., Hagiwara, A., et al., 1993. Morpho-physiological function and role of omental milky spots as omentum-associated lymphoid tissue (OALT) in the peritoneal cavity. Lymphology 26 (2), 90–101.

Siegel, R.L., Miller, K.D., Jemal, A., 2020. Cancer statistics, 2020. CA Cancer J. Clin. 70 (1), 7–30.

Sopo, M., Anttila, M., Hamalainen, K., Kivela, A., Yla-Herttuala, S., Kosma, V.M., et al., 2019. Expression profiles of VEGF-A, VEGF-D and VEGFR1 are higher in distant metastases than in matched primary high grade epithelial ovarian cancer. BMC Cancer 19 (1), 584.

Sorensen, E.W., Gerber, S.A., Sedlacek, A.L., Rybalko, V.Y., Chan, W.M., Lord, E.M., 2009. Omental immune aggregates and tumor metastasis within the peritoneal cavity. Immunol. Res. 45 (2–3), 185–194.

Sung, H., Ferlay, J., Siegel, R.L., Laversanne, M., Soerjomataram, I., Jemal, A., et al., 2021. Global cancer statistics 2020: GLOBOCAN estimates of incidence and mortality worldwide for 36 cancers in 185 countries. CA Cancer J. Clin. 71 (3), 209–249.

Tang, H., Chu, Y., Huang, Z., Cai, J., Wang, Z., 2020. The metastatic phenotype shift toward myofibroblast of adipose-derived mesenchymal stem cells promotes ovarian cancer progression. Carcinogenesis 41 (2), 182–193.

van Baal, J., van Noorden, C.J.F., Nieuwland, R., Van de Vijver, K.K., Sturk, A., van Driel, W.J., et al., 2018. Development of peritoneal carcinomatosis in epithelial ovarian cancer: a review. J. Histochem. Cytochem. 66 (2), 67–83.

Vitale, I., Manic, G., Coussens, L.M., Kroemer, G., Galluzzi, L., 2019. Macrophages and metabolism in the tumor microenvironment. Cell Metab. 30 (1), 36–50.

Wei, M., Yang, T., Chen, X., Wu, Y., Deng, X., He, W., et al., 2017. Malignant ascites-derived exosomes promote proliferation and induce carcinoma-associated fibroblasts transition in peritoneal mesothelial cells. Oncotarget 8 (26), 42262–42271.

Whitwell, H.J., Worthington, J., Blyuss, O., Gentry-Maharaj, A., Ryan, A., Gunu, R., et al., 2020. Improved early detection of ovarian cancer using longitudinal multimarker models. Br. J. Cancer 122 (6), 847–856.

Worzfeld, T., Pogge von Strandmann, E., Huber, M., Adhikary, T., Wagner, U., Reinartz, S., et al., 2017. The unique molecular and cellular microenvironment of ovarian cancer. Front. Oncol. 7, 24.

Yamamura, S., Matsumura, N., Mandai, M., Huang, Z., Oura, T., Baba, T., et al., 2012. The activated transforming growth factor-beta signaling pathway in peritoneal metastases is a potential therapeutic target in ovarian cancer. Int. J. Cancer 130 (1), 20–28.

Yeung, T.L., Leung, C.S., Yip, K.P., Au Yeung, C.L., Wong, S.T., Mok, S.C., 2015. Cellular and molecular processes in ovarian cancer metastasis. A review in the theme: cell and molecular processes in cancer metastasis. Am. J. Physiol. Cell Physiol. 309 (7), C444–C456.

Zhang, Y., Tang, H., Cai, J., Zhang, T., Guo, J., Feng, D., et al., 2011. Ovarian cancer-associated fibroblasts contribute to epithelial ovarian carcinoma metastasis by promoting angiogenesis, lymphangiogenesis and tumor cell invasion. Cancer Lett. 303 (1), 47–55.

Zhao, Y., Chen, C., Xu, X., Ge, X., Ding, K., Zheng, S., et al., 2021. An efficient prognostic immune scoring system for colorectal cancer patients with peritoneal metastasis. Onco. Targets. Ther. 10 (1), 1901464.

FSC
www.fsc.org
MIX
Papier | Fördert gute Waldnutzung
FSC® C083411

Kolibri 360 GmbH
Ferdinand-Jühlke-Straße 7
99095 Erfurt, Deutschland
produktsicherheit@kolibri360.de